Springer Theses

Recognizing Outstanding Ph.D. Research

For further volumes:
http://www.springer.com/series/8790

Aims and Scope

The series “Springer Theses” brings together a selection of the very best Ph.D. theses from around the world and across the physical sciences. Nominated and endorsed by two recognized specialists, each published volume has been selected for its scientific excellence and the high impact of its contents for the pertinent field of research. For greater accessibility to non-specialists, the published versions include an extended introduction, as well as a foreword by the student’s supervisor explaining the special relevance of the work for the field. As a whole, the series will provide a valuable resource both for newcomers to the research fields described, and for other scientists seeking detailed background information on special questions. Finally, it provides an accredited documentation of the valuable contributions made by today’s younger generation of scientists.

Theses are accepted into the series by invited nomination only and must fulfill all of the following criteria

- They must be written in good English.
- The topic should fall within the confines of Chemistry, Physics and related interdisciplinary fields such as Materials, Nanoscience, Chemical Engineering, Complex Systems and Biophysics.
- The work reported in the thesis must represent a significant scientific advance.
- If the thesis includes previously published material, permission to reproduce this must be gained from the respective copyright holder.
- They must have been examined and passed during the 12 months prior to nomination.
- Each thesis should include a foreword by the supervisor outlining the significance of its content.
- The theses should have a clearly defined structure including an introduction accessible to scientists not expert in that particular field.

Nele Boelaert

Dijet Angular Distributions in Proton–Proton Collisions

At $\sqrt{s}$ = 7 TeV and $\sqrt{s}$ = 14 TeV

Doctoral Thesis accepted by
Lund University, Sweden

Springer

Author
Dr. Nele Boelaert
Department of Physics
Lund University
22100 Lund
Sweden
e-mail: nele.boelaert@hep.lu.se

Supervisor
Prof. Torsten Åkesson
Department of Physics
Lund University
22100 Lund
Sweden
e-mail: Torsten.Akesson@hep.lu.se

ISSN 2190-5053
e-ISSN 2190-5061
ISBN 978-3-642-26993-6
ISBN 978-3-642-24597-8 (eBook)
DOI 10.1007/978-3-642-24597-8
Springer Heidelberg Dordrecht London New York

Softcover reprint of the hardcover 1st edition 2012

Cover design: eStudio Calamar, Berlin/Figueres

Printed on acid-free paper

Springer is part of Springer Science+Business Media (www.springer.com)

Supervisor's Foreword

One of the most common experimental methods in Physics is to illuminate a target, observe the pattern of the scattering, and learn something about the structure of the target. A famous example is of course the experiment of Rutherford in 1911, when he illuminated a gold foil with alpha particles and discovered the atomic nucleus. A later example is the electron scattering experiment using the Stanford Linear Accelerator which led to the discovery of the quarks inside the protons and neutrons. The highest resolution is obtained with the highest energy particles, and today, those are to be found at the Large Hadron Collider (LHC) at CERN. In the historic examples above, the targets were at rest in the laboratory, while the scattering is in a different rest-frame at a collider. Nevertheless, the principle is the same. By studying the angular distribution of the scattered quarks and gluons, at the LHC, their possible structure and the forces acting between them, are explored. This was the thesis topic of Dr Boelaert.

In 2006 the Physics Department at Lund University got the opportunity to launch a graduate school in High Energy Physics, thanks to a Marie Curie grant from the 6th framework programme of the European Union. It allowed seven students to start as graduate students. The students came from Belgium, China, Denmark, Italy, Romania and UK, and studied and did their research together in Lund during four years. I believe that this has given them and us, the permanent staff, an experience for life. Nele Boelaert came to this group with a degree in Engineering Physics from the Technical University in Gent/Belgium and I had the privilege to become her supervisor.

My interactions with her were a sequence of positive surprises in learning how much she had done, the quality of what she had done, and her innovative approach.

N. Boelaert started before the LHC became operational, and the original plan was that after a year of courses and a year of research preparation, she would have a few years of data from the LHC as basis for her thesis. However, like many frontier projects, the LHC encountered some delays. This gave her time to enter deeper into the phenomenological aspects of the measurements to be done, but also reduced the prospects of the scientific potential of the data that would be at hand for the thesis. Nevertheless, the LHC started during the last year of her position in

Lund allowing her to get results from the first, but limited, data from this new facility. This data was just a tiny first step into the vast space that the coming years of LHC operation will allow to be explored.

Her analysis has shown how well these measurements can be done, how well our current theory, the Standard Model, describes the entering point into the space opened up the LHC, and has shown the precision of the understanding of ATLAS at the LHC that has been built up by the scientific collaboration that constructed this facility. Nele Boelaert's thesis has set the best possible stage to pursue this exploration in the years to come as the LHC integrates luminosity and increases its energy. Her thesis is very well structured and written, and provides an excellent introduction for new graduate students to come.

Lund, August 2011 Torsten Åkesson

Acknowledgements

I would not have been able to complete this thesis without the help and influence of many people.

First of all, I wish to thank my main supervisor Torsten, because his suggestion to work on dijet angular distributions has resulted into a very fascinating and versatile research project.

Else became my co-supervisor nearly two years ago. Her advice concerning my research and work in the ATLAS collaboration has been extremely helpful. Else has given me the necessary confidence to grow and stand up inside the collaboration.

My warmest gratitude goes to Torbjörn. As my theoretical co-supervisor, he has led me successfully through my first phenomenology study. But his help has gone far beyond that; he has always been eager to share his very broad experience with me, explaining me everything I wanted to know, and giving me insight in how to proceed with my work. I have taken up so many hours of his time without him showing the least bit of impatience, and for all this I am very grateful.

I also wish to thank everybody who has been involved in setting up and organizing the Lund-HEP EST graduate school, which has given me the opportunity to start a PhD in high energy physics.

Part of my work consisted of using a Monte Carlo particle generator which was created by Leif. I appreciate very much the time he took for helping me use it and interpret the results.

Furthermore, I want to thank the ATLAS collaboration for integrating me and giving me scope to develop and learn. In particular, I wish to thank those people I have closely worked with on the dijet analysis: PO, Frederik, Lukas, Georgios and Sing. Our group has been very successful, with results of the measurements being published only a few months after the data had been recorded.

My work on ATLAS data quality was under the supervision of Michael, and I wish to express my gratitude to him because he has taught me well in the field of software development and the ATLAS offline framework.

Here in Lund, I wish to thank Anders, the head of the division of Experimental High-Energy Physics, because he has given me excellent feedback on my thesis.

Not only was he a very good listener to any kind of problem but he also always did whatever he could to help solving them.

Nowadays computers cannot be left out when studying high energy physics. Even though they are such a great aid, they often don't do what you want them to do. Fortunately, whenever I had a problem that seemed unsolvable, Richard would help me out. But more importantly, as my significant other, he has cared for me in a much broader context.

Finally I wish to thank my parents. They have been a constant and great support throughout my whole life. Without them, I wouldn't even have started this PhD study.

This thesis work was partly financed from the Marie Curie Mobility-2 action of the European Union 6th framework programme. Swedish participation in ATLAS was granted by the Swedish Research Council (VR) and the Knut and Alice Wallenberg foundation (KAW).

Abstract

Dijet angular distributions provide an excellent tool for looking at high transverse momentum parton interactions in order to study both QCD and new physics processes. With the Large Hadron Collider (LHC) recently brought into use, an unprecedented energy regime has opened up. ATLAS is one of the experiments at the LHC. Its high performance calorimeter system providing near hermetic coverage in the pseudorapidity range $|\eta| < 4.9$, enables ATLAS to perform reliable jet measurements. Detailed Monte Carlo studies at $\sqrt{s}$ = 14 TeV, the LHC design collision energy, and at $\sqrt{s}$ = 7 TeV, the collision energy foreseen for the initial years of the LHC operation, clearly show that dijet angular distributions can be used to discriminate the Standard Model from a new physics model describing gravitational scattering and black hole formation in large extra dimensions. When considering only the shape of the distributions, both the theoretical and the experimental uncertainties are predicted to be small in those regions where new physics is expected to show up. The study at $\sqrt{s}$ = 7 TeV indicates that ATLAS is already sensitive to large extra-dimensional gravity mediated effects with 1 pb^{-1} of data. The first measurement of dijet angular distributions at $\sqrt{s}$ = 7 TeV with ATLAS was carried out in two mass bins, using data that were recorded early 2010, corresponding to an integrated luminosity of about 15 nb^{-1}. The measurement shows good agreement with QCD predictions and demonstrates that ATLAS is ready to search for new physics in the dijet angular distributions with more data.

List of Publications Related to this Thesis

1. Dijet angular distributions at $\sqrt{s} = 14$ TeV
   By N. Boelaert
   Published in Proceedings of Science (EPS-HEP 2009) 298
2. Software design for prompt assessment of time-varying data quality
   By N. Boelaert, M. D'Onofrio, A. Dotti, C. Guyot, M. Hauschild, R. Hawkings, B. Heinemann, A. Höcker, V. Kazazakis, E. Lytken, M. Martínez-Perez, R. McPherson, P. U. E. Onyisi, A. Schaetzel, R. Seuster and M. G. Wilson
   Published as ATLAS internal note, ATL-COM-GEN-2010-002-Geneva: CERN, 2010
3. Implementation of the GravADD generator in Athena
   By N. Boelaert
   Published as ATLAS internal note, ATL-PHYS-INT-2010-012-Geneva: CERN, 2010
4. Dijet angular distributions at $\sqrt{s} = 14$ TeV
   By N. Boelaert and T. Åkesson
   arXiv:0905.3961 [hep-ph]
   Published in The European Physical Journal C, 33, (2010) 343–357
5. ATLAS sensitivity to contact interactions and large extra dimensions using dijet events a $\sqrt{s} = 7$ TeV
   By N. Boelaert, G. Choudalakis, P. O. Deviveiros, E. Feng, H. Li, J. Poveda, L. Pribyl, F. Ruehr and S. L. Wu
   Published as ATLAS internal note, ATL-COM-PHYS-2010-136-Geneva: CERN, 2010
6. High-pT dijet angular distributions in pp interactions at $\sqrt{s} = 7$ TeV measured with the ATLAS detector at the LHC
   By N. Boelaert, R. Buckingham, S. L. Cheung, G. Choudalakis, T. Davidek, P. O. De-Viveiros, E. Feng, J. Frost, M. Kaneda, H. Li, H. Peng, L. Pribyl, M. Shupe, K. Terashi, S. L. Wu
   Published as ATLAS internal note, ATL-COM-PHYS-2010-359-Geneva:

CERN, 2010
Presented at the International Conference on High Energy Physics 2010, Paris
Presented at LBNLMIT10, Cambridge, USA, a joint workshop between Lawrence Berkeley National Laboratory and Massachusetts Institute of Technology

Contents

Chapter 1
Introduction

1.1 The Standard Model and Beyond

Already in the sixth century BC, the Greeks believed that matter is composed of elementary particles. But it took until the discovery of the electron at the end of the ninetieth century to finally get a breakthrough in our understanding of the structure of matter, which forced the twentieth century into a rapid growth of theories and knowledge.

Nowadays scientists believe that the Universe is made of elementary particles which are governed by four fundamental forces: electromagnetism, strong and weak force, and gravitation [1]. The *Standard Model* of particle physics (SM) is a theory of elementary particles and all forces but gravity, which describes existing data very well [2].

The elementary particles of the SM are twelve fermions (and their antifermions), twelve gauge bosons and one neutral Higgs particle. Fermions are half-integer spin particles which respect the Pauli Exclusion Principle, but gauge bosons have integer spin and do not follow this rule. The Higgs particle is a scalar, meaning that it has spin 0.

The gauge bosons mediate the forces between the elementary particles. Each of the forces is associated with a charge: electric charge, weak charge and strong charge. The strong charge is also called the color charge, or color for short.

Two types of fundamental fermions exist: quarks and leptons. The leptons come in three lepton families: electron (ν_e, e), muon (ν_μ, μ), and tau (ν_τ, τ). They can also be classified according to their charge: the neutral neutrinos ν_e, ν_μ, ν_τ and the negatively charged e^-, μ^- and τ^-.

The quarks come in six flavors and, like the leptons, they can be grouped into three quark families: (u, d), (c, s) and (t, b). The u, c, t quarks have electric charge 2e/3, and the d, s, b quarks have charge $-e/3$, with e being the elementary electric charge.

Fermions can be decomposed into left-handed doublets and right-handed singlets of the electroweak force. The three fermion families can be summarized as follows:

N. Boelaert, *Dijet Angular Distributions in Proton–Proton Collisions*, Springer Theses,
DOI: 10.1007/978-3-642-24597-8_1, © Springer-Verlag Berlin Heidelberg 2012

Table 1.1 Nonzero fermion masses in the SM.

1st family		2 *st* family		3*st* family	
Name	Mass (MeV/c^2)	Name	Mass (GeV/c^2)	Name	Mass (GeV/c^2)
u	1.5–3.3	c	$1.27^{+0.07}_{-0.11}$	t	171.2 ± 2.1
d	3.5–6.0	s	$0.104^{+0.026}_{-0.034}$	b	$4.20^{+0.17}_{-0.07}$
e	0.511	μ	0.106	τ	1.777

$$\text{1st family:} \begin{pmatrix} \nu_e \\ e^- \end{pmatrix}_L, e^-_R, \begin{pmatrix} u \\ d \end{pmatrix}_L, u_R, d_R$$

$$\text{2nd family:} \begin{pmatrix} \nu_\mu \\ \mu^- \end{pmatrix}_L, \mu^-_R, \begin{pmatrix} c \\ s \end{pmatrix}_L, c_R, s_R$$

$$\text{3rd family:} \begin{pmatrix} \nu_\tau \\ \tau^- \end{pmatrix}_L, \tau^-_R, \begin{pmatrix} t \\ b \end{pmatrix}_L, t_R, b_R$$

Each fermion has an associated antifermion with the same mass, but opposite charges. Unlike leptons, quarks have color charge and engage in the strong interaction.

As for the fermion masses, the members of the first family are very light, but the masses increase with the family number, which explains why ordinary matter is made of the first family. According to the original formulation of the SM, neutrinos are massless, but nowadays it is generally accepted that they have small but non-vanishing mass. A summary of quark and lepton masses is given in Table 1.1. Quarks are confined (more about confinement in Chap. 2), and therefore their masses cannot be directly measured, which explains the rather large error ranges in the table. On the other hand, the masses of the charged leptons are known to a high precision. The SM is a quantum field theory which is based on the gauge symmetry $SU(3)_C \times SU(2)_L \times U(1)_Y$. This gauge group includes the symmetry group of the strong interactions $SU(3)_C$—the subscript C stands for color—and the symmetry group of the electroweak interactions $SU(2)_L \times U(1)_Y$, where the subscript L indicates that among fermions only left-handed states transform nontrivially under the electroweak $SU(2)$, and the Y stands for hypercharge, the generator of $U(1)$. The group symmetry of the electromagnetic interactions, $U(1)_{em}$, appears in the SM as a subgroup of $SU(2)_L \times U(1)_Y$ and it is in this sense that the weak and electromagnetic interactions are said to be unified.

The gauge sector of the SM is summarized in Table 1.2 and is composed of eight gluons which are the gauge bosons of $SU(3)_C$, and the γ, $W^\pm$ and Z particles which are the four gauge bosons of $SU(2)_L \times U(1)_Y$. The gluons are massless and electrically neutral, and carry color quantum number. There are eight gluons with different color-anticolor combinations. The consequence of the gluons carrying color is that they interact not only with the quarks but also with other gluons. The interactions of quarks and gluons can be described by a theory called quantum chromodynamics, or QCD in short, and we will discuss this in detail in the next chapter.

Table 1.2 Gauge bosons in the SM.

Force	Gauge boson	Mass(GeV/c^2)	Gauge group	Charges and range
Electromagnetism	Photon (γ)	0	Unbroken $U(1)$ combination of $SU(2) \times U(1)$	γ is electrically neutral, force has infinite range
Weak force	$W^{\pm} Z$	80.4 91.2	Broken combination of $SU(2) \times U(1)$	W and Z have weak charge, W has electric charge, interaction is short ranged
Strong (or color) force	8 gluons (g)	0	$SU(3)_C$	Gluon carries color force force has finite range

The weak bosons, $W^{\pm}$ and Z are massive particles and also self-interacting. The $W^{\pm}$ bosons are electrically charged with $Q_e = \pm e$ respectively and the Z boson is electrically neutral. The photon γ is massless, chargeless and non-selfinteracting as it does not carry electric charge. Of all gauge bosons, only the gluons carry color.

The $W^{\pm}$ and Z bosons are heavy, of the order of 100 GeV, which implies that the weak force is short ranged (of the order of 10^{-3}fm). Because the photon has zero mass, the electromagnetic force has infinite range. The gluons are massless too, but because they are self interacting, the range of the strong force is limited to distances <1fm.

As for the strength of the three interactions, the electromagnetic interactions are governed by the magnitude of the electromagnetic coupling constant e or equivalently $\alpha = \frac{e^2}{4\pi}$, which at low energies is given by the fine structure constant, $\alpha(Q = m_e) = \frac{1}{137}$. The weak interactions at energies much lower than the exchanged gauge boson mass, have an effective (weak) strength given by the Fermi constant $G_F = 1.167 \times 10^{-5}\ \mathrm{GeV}^{-2}$. The name of strong interactions is due to their comparatively larger strength than the other interactions. This strength is governed by the size of the strong coupling constant g_S or equivalently $\alpha_S = \frac{g_S^2}{4\pi}$ which is about ~ 1 at energies comparable to hadron masses.

The fact that the weak gauge bosons are massive particles indicates that $SU(3)_C \times SU(2)_L \times U(1)_Y$ is not a symmetry of the vacuum. On the other hand, the fact that the photon is massless reflects that $U(1)_{em}$ is a good symmetry of the vacuum. We therefore expect spontaneous symmetry breaking in the SM and it must occur in the following way:

$$SU(3)_C \times SU(2)_L \times U(1)_Y \to SU(3)_C \times U(1)_{em} \tag{1.1}$$

The above pattern is implemented in the SM by means of the so-called Higgs mechanism, which provides the proper masses to the $W^{\pm}$ and Z gauge bosons and to the fermions, and leaves as a consequence the prediction of a new particle: the Higgs boson particle, which must be a scalar and electrically neutral. This particle has not been observed so far.

Although the SM succeeds at describing precisely phenomena in the GeV energy range, there are a variety of indications that more fundamental physics remains to be discovered. For example the experimental observations that neutrinos have mass [3], are not quite compatible with the original formulation of the SM. Also the hierarchy problem—fine tuning problems from radiative corrections to the Higgs mass [4]—requires physics beyond the SM. New physics will extend and strengthen the foundations of the SM, but the SM will remain a valid effective description within its energy range, whether it is a fundamental theory or not.

A large number of physics scenarios beyond the SM have been considered. One way to extend the SM, is to try for further unification by constructing models that unify quarks and leptons, and the electroweak and strong force. These theories are called Grand Unified Theories.

Supersymmetry is another possibility. This is the name given to a hypothetical symmetry of nature which relates fermions and bosons. Every particle (quark, lepton and boson) has a superpartner in this theory. Because supersymmetry can only exist as a broken theory in nature, all superpartners have a high mass, which explains why they have not been observed yet.

Exploring the world in extra dimensions is another way of dealing with unexplained phenomena. The ADD model addresses the hierarchy problem by assuming the existence of large extra dimensions in which gravity is allowed to propagate, while the SM fields are confined to a four-dimensional membrane. More about this theory in Chap. 5.

1.2 High Energy Physics Experiments and the Large Hadron Collider

It might sound contradictory, but experiments carried out for testing the smallest particles physicists believe exist, are typically huge in terms of design and number of participating scientists. Particle accelerator experiments belong to that category; particles are brought to almost the speed of light and are then directed into collisions with particles traveling in the opposite direction.

The Large Hadron Collidor (LHC) [5] is the world's largest and most powerful particle accelerator, designed to either collide protons at an energy of 14 TeV per particle pair, or lead nuclei at an energy of 5.5 TeV per nucleon pair. However, in the early phase of LHC (2009–2012), only protons with an energy 3.5 TeV are brought to collide. The accelerator was built by the European Organization for Nuclear Research (CERN) nearby Geneva, with the intention of testing various aspects of high energy

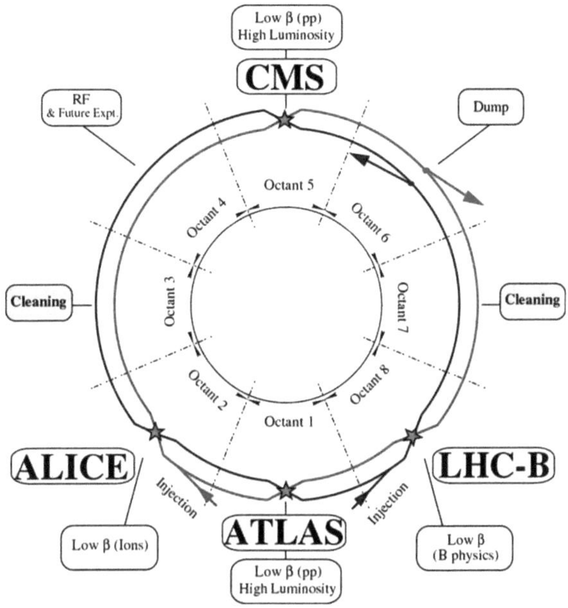

Fig. 1.1 Schematic view of the LHC accelerator [6]

physics, ranging from more precise measurements of Standard Model parameters to the search for new physics phenomena and properties of strongly interacting matter at extreme energy density.

Most of the accelerator is situated in a 27 km circular tunnel underground. A salient feature of the LHC is the superconducting helium cooled dipole magnet system which operates at 8.3 T in order to keep 7 TeV protons in their circular orbits.

The LHC design luminosity for proton collisions is $10^{34}\text{cm}^{-2}s^{-1}$, which will only be reached after an initial period of running at lower luminosity. In a proton–proton run, the LHC beam is subdivided into bunches with a spacing of 25 ns or 7.5 m. At design luminosity, 2808 bunches will circulate in the ring, each bunch containing about 10^{11} particles, and the total energy of the beam will be around 362 MJ.

Four large detectors are placed at different interaction points around the ring, see Fig. 1.1; ATLAS [7] (A Toroidal LHC ApparatuS), CMS [8] (Compact Muon Solenoid), ALICE [9] (A Large Ion Collider Experiment) and LHCb [10] (Large Hadron Collider beauty). ATLAS and CMS are general purpose detectors, designed to observe at LHC design luminosity phenomena that involve highly massive particles which were not observable using earlier lower-energy accelerators and might shed light on new theories of particle physics beyond the Standard Model. ALICE is designed for heavy ion collisions and aims to study a “liquid” form of matter called quark-gluon plasma that existed shortly after the Big Bang. LHCb is a specialized b-physics experiment, particularly aimed at measuring the parameters of CP violation [11] in the interactions of b-hadrons.

The LHC became operational at the end of 2009. After a few months of commissioning at lower energies, the LHC was brought to collide protons at a center of mass energy of 7 TeV early 2010. A further increase in energy is not foreseen for 2010/2011. By the end of June 2010, an integrated luminosity of nearly 30 nb^{-1} was recorded by ATLAS. At present, the instantaneous luminosity is increasing exponentially and the goal for 2010 is to record about 100 pb^{-1} of data.

1.3 This Thesis: Dijet Angular Distributions at LHC Energies

The QCD production of jets of particles is by far the most dominant hard process at colliders. Previous experiments—at lower energies than the LHC—have shown that the study of the angular correlation between the two strongest (hardest) jets provides a good tool for probing both the Standard Model and new physics theories.

This thesis continues this study at LHC energies. The aim is to carry out the measurement of dijet angular distributions with ATLAS and compare the results with theory predictions. Both the Standard Model and a new physics model for large extra dimensions have been considered.

Due to the delay of the LHC startup, this thesis is largely based on Monte Carlo simulations of the collisions. The simulated collisions have been used for studying both the phenomenology of the collision processes and how they show up in the acquired data and can be analyzed. But besides these detailed Monte Carlo studies, we will show the first results of the actual measurement as well, and make the comparison with the Standard Model.

The first chapters (Chaps. 2–5) of this thesis mainly contain theory. An introduction to QCD and collider physics is given in Chap. 2, which is followed by a detailed discussion about next-to-leading order Monte Carlo techniques in Chap. 3. Dijet physics, with emphasis on dijet angular distributions, is discussed in Chap. 4. Chapter 5 is about new physics, more precisely about gravitational scattering and black hole production in a world with large extra dimensions. This chapter also details the implementation of a dedicated Monte Carlo particle generator in the ATLAS software framework, which was done in order to be able to compare official ATLAS (simulated) data with the new physics predictions. The theory chapters are followed by three chapters about the ATLAS detector. Chapter 6 gives a general description of the detector, together with a rather detailed report about the data quality work done during this PhD study. Chapters 7 and 8 focus on the jet reconstruction; Chapters 7 reviews the methods for the jet reconstruction and its performance evaluation based on Monte Carlo simulated data, while Chap. 8 focuses on the jet reconstruction using ATLAS data that were recorded at the end of 2009 and in the first half of 2010.

A detailed Monte Carlo phenomenology study at a center of mass energy of $\sqrt{s} = 14$ TeV is presented in Chap. 9. Both QCD and new physics coming from gravity mediated effects in large extra dimensions, are topics of discussion.

Since in the initial phase of LHC, protons only collide at half their nominal energy, ATLAS has performed its first measurement of dijet angular distributions at $\sqrt{s} = 7$ TeV. Chapter 10 contains a dedicated Monte Carlo study aimed at preparing the ATLAS detector for this measurement. Apart from a phenomenology study, this chapter also investigates the technical aspects of the measurement and the ATLAS sensitivity to new physics addressable by dijet studies.

The measurement of dijet angular distributions is presented in Chap. 11, and the comparison with Standard Model predictions is made.

Finally the conclusions of this thesis and outlook are presented in Chap. 12.

1.4 Author's Contribution

The work in this thesis was published in a number of publications. Below I will make a chronological listing of my research activities during the past four years, and describe my contributions to scientific papers.

Every PhD student in the ATLAS collaboration is expected to work on a technical task, and mine was situated in the ATLAS data quality framework; during the second year of my PhD, I worked on a web display for prompt monitoring of the data quality of reconstructed data in ATLAS. More specifically, I developed the handi and DQWebDisplay applications for the ATLAS data quality group, and I assisted the ATLAS physics validation group with using the tools to set up their own display, PhyValMon.

The work was published in Ref. [12] and is also summarized in Sect. 6.4 of this thesis. Writing the technical note was my initiative and I was the main responsible. But I worked closely with most of the authors and only about 50% of the text was written by me.

On September 10 2008, the LHC accelerated its first protons ever, but shortly after that—on September 19—a major accident happened that forced the LHC to be in repair for over one year (until November 2009).

Because of the delay, I decided to move away from pure experimental work and focus on the phenomenology of dijet angular distributions at $\sqrt{s} = 14$ TeV. First I studied the distributions in the context of QCD and later on, I included new physics from large extra dimensions. All details can be found in Chap. 9.

A publication concerning this study was written mainly by me, and was eventually published [13]. In the summer of 2009 I gave a talk about this study at the 2009 Europhysics Conference on High Energy Physics in Krakow, Poland [14].

In the beginning of 2009, it became clear that the LHC was initially going to collide protons at $\sqrt{s} = 7$ TeV, and that collisions were foreseen by the end of the year.

We therefore set up a small team within the ATLAS collaboration that worked on the preparation of the early measurement of dijet angular distributions. QCD and many other physics models were investigated, but my focus was on QCD and new physics from large extra dimensions only. I was able to use my experience from my previous study to work on the phenomenology part, but I also investigated the influence of pure detector effects, such as jet energy calibration.

An ATLAS internal note was published in Ref. [15]. My contributions concern the selection cuts in kinematics, the QCD theory calculations, the estimate of theoretical uncertainties and NLO effects, and the sensitivity to gravitational scattering and semiclassical black holes in large extra dimensions. Chapter 10 is entirely devoted to this study.

In order to be able to study gravitational scattering and semiclassical black hole formation with ATLAS, I needed to implement a dedicated standalone Monte Carlo generator, called GravADD, in the ATLAS software framework. This is a non-trivial task since the ATLAS software framework has specific requirements and conventions. In order to provide documentation for future ATLAS users and programmers, I published an internal note [16]. Details are given in Sect. 5.5.4.

On March 30, 2010. The LHC started to collide protons at $\sqrt{s} = 7\,\text{TeV}$. The small team that worked meticulously on the preparation of the measurement, was now able to actually perform the measurement in a rather limited amount of time. I worked on the Standard Model theory predictions and uncertainties and wrote the code to run over the data and create the final plots. More about this in chapter 11. The analysis was made public in July 2010 at the International Conference on High Energy Physics in Paris [17]. In August 2010, I presented the results at LBNLMIT10, a joint Berkeley-MIT workshop on Implications of First LHC Data in Cambridge, USA.

References

1. B. Povh, K. Rith, C. Scholz, F. Zetsche, *Particles and Nuclei* (Springer, Heidelberg, 2003)
2. J.F. Donoghue, E. Golowich, B.R. Holstein, *Dynamics of the Standard Model Cambridge Monographs on Particle Physics Nuclear Physics and Cosmology* (Cambridge University Press, Cambridge, 1996)
3. Super-Kamiokande Collaboration: Y. Fukuda, et al., Evidence for oscillation of atmospheric neutrinos. Phys. Rev. Lett **81**(8) 1562–1567 (1998)
4. S.P. Martin, A Supersymmetry Primer, (1997) hep-ph/9709356
5. O.S. Bruning et al., LHC design report. Vol. I: The LHC main ring. CERN. CERN-2004-003-V-1. (2004)
6. LHC Machine Outreach, http://lhc-machine-outreach.web.cern.ch/lhc-machine-outreach.
7. W.W. Armstrong, et al., ATLAS: Technical proposal for a general-purpose pp experiment at the Large Hadron Collider at CERN. CERN. CERN-LHCC-94-43 (1994)
8. CMS Collaboration. CMS, the compact muon solenoid: technical proposal. CERN, Geneva (1994)
9. ALICE Collaboration. ALICE: Technical proposal for a large ion collider experiment at the CERN LHC. CERN, Geneva (1995)
10. LHCb Collaboration LHCb: Technical Proposal. CERN, Geneva (1998)
11. C. Amsler et al., Review of particle physics. Phys. Lett. **B667**, 1 (2008)
12. N. Boelaert, et al., Software design for prompt assessment of time-varying data quality. Technical Report ATL-COM-GEN-2010-002, CERN, Geneva (2010)

13. N. Boelaert, T. Åkesson, Dijet angular distributions at $\sqrt{s} = 14\ TeV$. EPJ C **66**, 343–357 (2010) hep-ph/09053961
14. N. Boelaert, Dijet angular distributions at $\sqrt{s} = 14$ TeV, in *EPS-HEP. Proceedings of science* (2009)
15. N. Boelaert, G. Choudalakis, P.O. Deviveiros, E. Feng, H. Li, J. Poveda, L. Pribyl, F. Ruehr, S.L. Wu, ATLAS sensitivity to contact interactions and gravity mediated effects in large extra dimensions using dijet events at $\sqrt{s} = 7$ TeV. Technical Report ATL-COM-PHYS-2010-136, CERN, Geneva (2010)
16. N. Boelaert, Implementation of the GravADD generator in Athena. Technical Report ATL-PHYS-INT-2010-012, CERN, Geneva (2010)
17. N. Boelaert, R. Buckingham, S.L. Cheung, G. Choudalakis, T. Davidek, P.O. Deviveiros, E. Feng, M. Kaneda, J. Frost, H. Li, H. Peng, L. Pribyl, F. Ruehr, M. Shupe, K. Terashi, S.L. Wu, High-pT dijet angular distributions in pp interactions at $\sqrt{s} = 7$ TeV measured with the ATLAS detector at the LHC. Technical Report ATL-COM-PHYS-2010-359, CERN, Geneva (2010)

Chapter 2
Introduction to QCD and Collider Physics

2.1 Quantum Chromodynamics (QCD)

Quantum chromodynamics (QCD) is the theory of the strong interaction, describing the interactions of the quarks and gluons, using the $SU(3)$ non-Abelian gauge theory of color charge [1]. The expression for the classical QCD Lagrangian density is given by:

$$\mathcal{L} = -\frac{1}{4} F^{A}_{\alpha\beta} F^{\alpha\beta}_{A} + \sum_{\text{flavors}} \bar{q}_a (i\gamma_\mu D^\mu - m)_{ab} q_b, \tag{2.1}$$

where the sum runs over the n_f different flavors of quarks ($n_f = 6$ in the SM), and α, β, γ are Lorentz indices. Throughout this entire chapter, we will work with the convention that repeated indices are implicitly summed over. $F^{A}_{\alpha\beta}$ is the field strength tensor derived from the gluon field $\mathcal{A}^{\mathrm{A}}_{\alpha}$:

$$F^{A}_{\alpha\beta} = [\partial_\alpha \mathcal{A}^{A}_{\beta} - \partial_\beta \mathcal{A}^{A}_{\alpha} - g_s f^{ABC} \mathcal{A}^{B}_{\alpha} \mathcal{A}^{C}_{\beta}] \tag{2.2}$$

The capital indices A, B and C run over the eight degrees of freedom of the gluon field. Note that it is the third (non-Abelian) term in the above expression that makes the gluons have self-interactions. This means that, unlike the photon in QED, the carrier of the color force is itself colored, a property that is giving rise to asymptotic freedom (see further in the text). The numbers f^{ABC} are structure constants of the $SU(3)$ group. Quark fields q_a ($a = 1, 2, 3$) are in triplet color representation, with colors red (r), green (g) and blue (b).

The strong coupling strength g_s in Eq. 2.2 is used to define the strong coupling constant $\alpha_s = g_s^2/4\pi$. D in Eq. 2.1 stands for the covariant derivative, which takes, acting on triplet and octet fields respectively, the form:

$$(D_\alpha)_{ab} = \partial_\alpha \delta_{ab} + ig(t^C \mathcal{A}^{C}_{\alpha})_{ab} \tag{2.3}$$

N. Boelaert, *Dijet Angular Distributions in Proton–Proton Collisions*, Springer Theses,
DOI: 10.1007/978-3-642-24597-8_2, © Springer-Verlag Berlin Heidelberg 2012

$$(D_\alpha)_{AB} = \partial_\alpha \delta_{\mathrm{AB}} + ig(T^C \mathcal{A}_\alpha^C)_{AB}, \tag{2.4}$$

where t and T are matrices in the fundamental and adjoint representations of $SU(3)$ respectively.

2.1.1 Perturbative QCD (pQCD)

By adding a gauge-fixing term to the classical QCD Lagrangian (Eq. 2.1):

$$\mathcal{L}_{\mathrm{gauge-fixing}} = -\frac{1}{2\lambda}(\partial^\alpha \mathcal{A}_\alpha^A)^2, \tag{2.5}$$

and a so-called ghost Lagrangian which is derived from a complex scalar field η^A and is needed because the theory is non-Abelian:

$$\mathcal{L}_{\mathrm{ghost}} = \partial_\alpha \eta^{A\dagger}(D^\alpha_{AB}\eta^B), \tag{2.6}$$

any process can be calculated in a perturbative way using Feynman rules which are obtained from replacing covariant derivatives by appropriate momenta. The Feynman rules in a covariant gauge are given in Fig. 2.1. However, a perturbative calculation generally requires 4-dimensional integrations over intermediate momentum states arising from gluon quantum fluctuations, which suffer from ultraviolet divergences.

A *renormalization* procedure is needed to remove these divergences, which essentially means that the Lagrangian is rewritten so that bare masses and coupling strengths are eliminated in favor of their physically measurable counterparts, giving rise to a renormalized Lagrangian [2]. Modified Feynman rules are derived from this Lagrangian and singularities in the contributions from individual diagrams are now absorbed by the physical quantities, leading to a finite result at the end.

Several renormalization methods are possible, and the exact definitions of physical quantities—masses and coupling constants—depend on the specific renormalization scheme used in the theory, but common to all schemes is the inclusion in the renormalized Lagrangian of a new, arbitrary parameter, with the dimension of mass, needed to define the physical quantities.This parameter is often called the renormalization scale μ_R. It appears in the intermediate parts of a calculation, but cannot ultimately influence the relations between physical observables.

A consequence of renormalization is that the definition of the physically observable quantities not only depends on μ_R, but also becomes scale dependent; when the theory is normalized at a scale μ_R but then applied to a very different scale Q (of the order of the momentum invariants of the reaction), the coupling constants and masses adjust to that scale, a process which is commonly referred to as the *running* of the coupling constants and masses.

The running of the coupling constant α_s is controlled by the β function [3], which is derived from the statement that a physical observable cannot depend on μ_R:

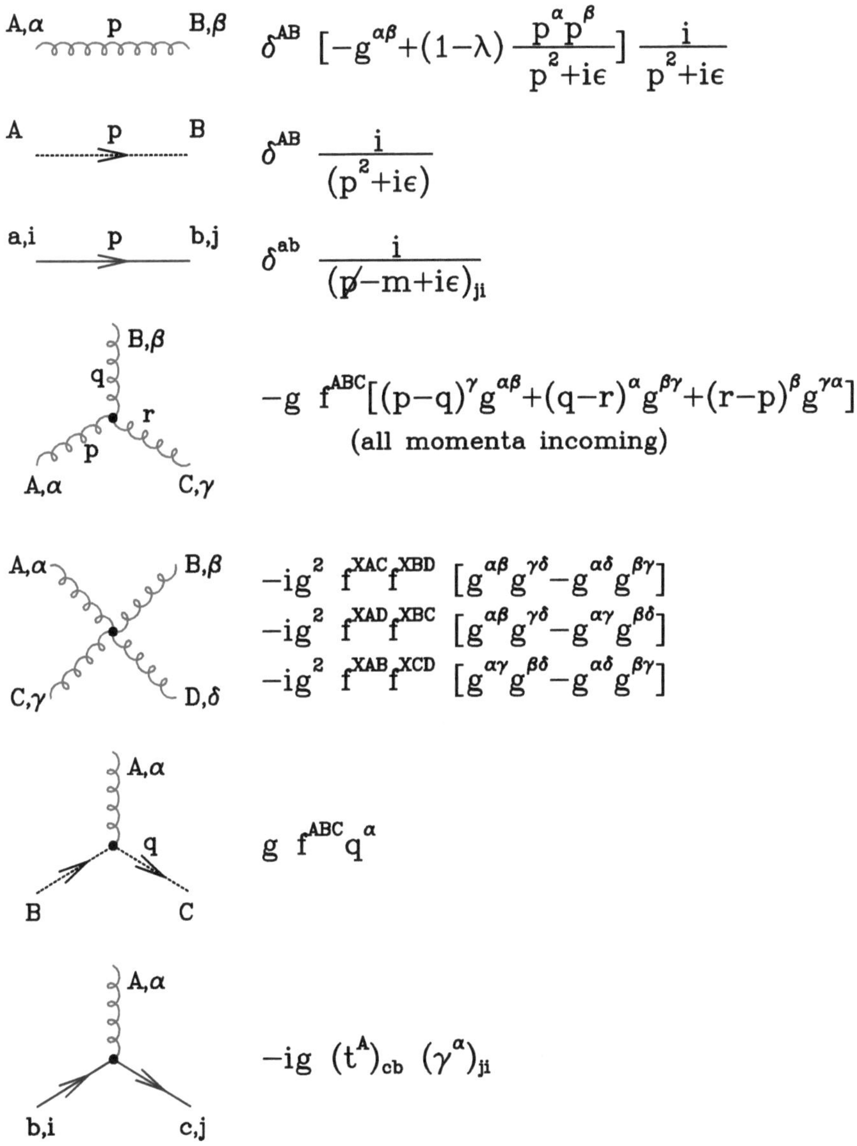

Fig. 2.1 Feynman rules for QCD in a covariant gauge from gluons (*curvy red lines*), fermions (*solid blue lines*) and ghosts (*dotted black lines*) [1]

$$Q\frac{\partial \alpha_s}{\partial Q} \equiv 2\beta_{\mathrm{QCD}} = -\frac{\beta_0}{2\pi}\alpha_s^2 - \frac{\beta_1}{4\pi^2}\alpha_s^3 - \mathcal{O}(\alpha_s^4), \tag{2.7}$$

with

$$\beta_0 = 11 - \frac{2}{3}n_f \tag{2.8}$$

$$\beta_1 = 51 - \frac{19}{3} n_f \tag{2.9}$$

Given that α_s is known (from experiment) at a certain scale Q_0, Eq. 2.7 can be used to calculate its value at any other scale Q:

$$\log(Q^2/Q_0^2) = \int_{\alpha_s(Q_0)}^{\alpha_s(Q)} \frac{d\alpha}{\beta(\alpha)} \tag{2.10}$$

Equation 2.10 is solvable using the leading-order (LO) term of $\beta(\alpha)$ only, which gives:

$$\alpha_s(Q) = \frac{\alpha_s(Q_0)}{1 + \frac{\beta_0}{2\pi}\alpha_s(Q_0)\ln(Q^2/Q_0^2)} \approx \alpha_s(Q_0)\Big(1 - \frac{\beta_0}{2\pi}\alpha_s(Q_0)\ln(Q^2/Q_0^2)\Big) \tag{2.11}$$

Another way to solve Eq. 2.7 is by introducing a dimensional parameter Λ, representing the mass scale at which α_s becomes infinite. This way, we get:

$$\alpha_s(Q) = \frac{4\pi}{\beta_0 \ln(Q^2/\Lambda^2)}\left[1 - \frac{2\beta_1}{\beta_0^2}\frac{\ln[\ln(Q^2/\Lambda^2)]}{\ln(Q^2/\Lambda^2)} + \mathcal{O}(\ln^{-2}(Q^2/\Lambda^2))\right] \tag{2.12}$$

Note that in Eqs. 2.11 and 2.12 the running of α_s with Q is logarithmic, so that we do not need to worry too much about choosing Q precisely.

Equation 2.12 illustrates the hallmark of QCD, namely *asymptotic freedom*: $\alpha_s \to 0$ as $Q \to \infty$. It also shows that QCD becomes strongly coupled at $Q \sim \Lambda$, which is at about 200 MeV. This implies that perturbative methods can be used in the short-distance limit, at scales Q much larger than Λ. The fact that the strong force becomes strong at larger distances, means that color charged particles cannot be isolated singularly and cannot be observed as states that propagate over macroscopic distances, a property which is called *confinement*. Only color singlet states composed of quarks and gluons, i.e. hadrons, can be observed. We will talk about hadronization in Sect. 2.5. Perturbative methods are no longer a valid approximation in this area.

Experiments usually report the strong coupling at the scale corresponding to the Z mass ($M_Z = 91.2$ GeV). The world average of $\alpha_s(M_Z)$ is determined from measurements which are based on QCD calculations in complete next-to-next-to leading order (NNLO) perturbation theory, giving $\alpha_s(M_Z) = 0.1182 \pm 0.0027$ [4].

2.2 The Parton Model

The high-energy interactions of hadrons are described by the QCD parton model [5, 6]. The basic idea of this model is that the hard scattering between two hadrons can be understood as the interaction between the partons—quarks and gluons with their masses neglected—that make up the hadrons.

A hadron consists of a number of valence quarks (e.g. *uud* for the proton) and an infinite sea of gluons and light quark-antiquark ($q\bar{q}$) pairs. The valence quarks carry the hadron's electric charge and baryon quantum numbers. When probed at a scale Q, the sea contains all quark flavors with mass $m_q \ll Q$. The gluons carry about 50% of the proton's total momentum. A *parton distribution function* (PDF) is used to denote the probability distribution that a quark, antiquark or gluon carries a given fraction of the momentum of the hadron.

The sea is not static, there is a continuous movement of gluons splitting and recombining into $q\bar{q}$ pairs, and both quarks and gluons can emit and absorb gluons as well. These processes imply that the transverse momenta of partons inside the hadron are not restricted to small values, and that the PDFs describing the partons depend on the scale Q that the hadron is probed with, a behavior which is known as a violation of Bjorken Scaling. At leading order, the dependence on Q is logarithmic.

If $q(x, Q)$ is the PDF describing quark Q, then $q(x, Q)\mathrm{d}x$ represents the probability that Q carries a momentum fraction between x and $x + \mathrm{d}x$ when the hadron is probed at a scale Q.

Each hadron has its own set of PDFs and separate PDFs are used for describing the sea and the valence quarks; the PDFs for the valence quarks are flavor specific, but QCD guarantees flavor number conservation of the sea quarks.

For example, for the proton at a scale of about 1 GeV, we can write:

$$u(x, Q) = u_v(x, Q) + u_s(x, Q) \tag{2.13}$$

$$d(x, Q) = d_v(x, Q) + d_s(x, Q) \tag{2.14}$$

Taking into account quark number conservation, the following sum rules apply:

$$\int_0^1 \mathrm{d}x\, u_v(x, Q) = 2 \tag{2.15}$$

$$\int_0^1 \mathrm{d}x\, d_v(x, Q) = 1 \tag{2.16}$$

And experimentally, it was found that:

$$\sum_q \int_0^1 \mathrm{d}x\, x[q(x, Q) + \bar{q}(x, Q)] \approx 0.5, \tag{2.17}$$

meaning that the quarks carry only about half of the proton's momentum (and the gluons the other half).

When a quark emits a gluon, it can acquire a large momentum k_T with probability proportional to $\alpha_s \mathrm{d}k_T^2/k_T^2$ at large k_T. This splitting diverges in the collinear region ($k_T \to 0$). This is not a physical divergence; it simply means that perturbative QCD is not a valid approximation in this region.

The way to solve this is to renormalize the PDFs by introducing a *factorization scale* μ_f. Similar to the renormalization scale, the factorization scale absorbs the divergences coming from interactions that are not calculable in perturbation theory. This way, the PDFs become scale dependent, just like the strong coupling constant discussed in the previous section.

Perturbative QCD carries no absolute prediction of the PDF, but does predict how the PDF scales with Q; these are the so called DGLAP (Dokshitzer–Gribov–Lipatov–Altarelli–Parisi) evolution equations [7–9]:

$$t\frac{\partial}{\partial t}\begin{pmatrix} q_i(x,t) \\ g(x,t) \end{pmatrix} = \frac{\alpha_s(t)}{2\pi}\sum_{q_i,\bar{q}_j}\int_x^1 \frac{\mathrm{d}\,z}{z}\begin{pmatrix} P_{q_j\to q_i g}(z,\alpha_s(t)) & P_{g\to q_i\bar{q}_i}(z,\alpha_s(t)) \\ P_{q_j\to g q_i}(z,\alpha_s(t)) & P_{g\to gg}(z,\alpha_s(t)) \end{pmatrix} \begin{pmatrix} q_j(x/z,t) \\ g(x/z,t) \end{pmatrix} \tag{2.18}$$

Here, $t = -Q^2$, $q_{i,j}(x,t)$ and $g(x,t)$ are the quark and gluon parton distribution functions respectively, and the functions $P_{a\to bc}(z)$ are the so called unregularized splitting kernels [1]. We will derive the DGLAP equations in Sect. 2.4

The DGLAP evolution equations specify the evolution of the parton density functions in the same way as the β function (Eq. 2.7) specifies the evolution of the strong coupling constant. When solving Eq. 2.18 to the leading order, the term $\partial t/t$ will cause the PDFs to obey a logarithmic dependence on $t = -Q^2$.

The DGLAP equations do allow for the evolution of the PDFs from a certain reference scale Q_0 onwards, but data are still needed to determine its value at the scale Q_0. Deep inelastic lepton-hadron scattering measurements are an excellent tool for probing PDFs and the reference scale is typically chosen around 1 GeV. Note that PDFs are universal, i.e. they can be determined from one type of experiment (e.g. $e^- p$ collisions) and used in another (e.g. pp collisions). In the past, leading-order matrix elements together with lowest order running of α_s (see Eq. 2.11) were used for the fit. Nowadays, also next-to-leading order (NLO) and even NNLO PDFs—resulting from a fit to NLO or NNLO matrix elements and a higher order running of α_s—have become available.

Historically there are two major collaborations working on PDFs: the CTEQ [10], and the MRST [11], nowadays MSTW [12], collaboration. Figure 2.2 shows the MRST2004NLO PDFs multiplied with x, for the up and down quark and the gluon inside the proton at $Q^2 = 10^4\ \mathrm{GeV}^2$. The gluon distribution is scaled with a factor 1/10 in order to fit into the plot. Note that the gluon distribution dominates at small values of x.

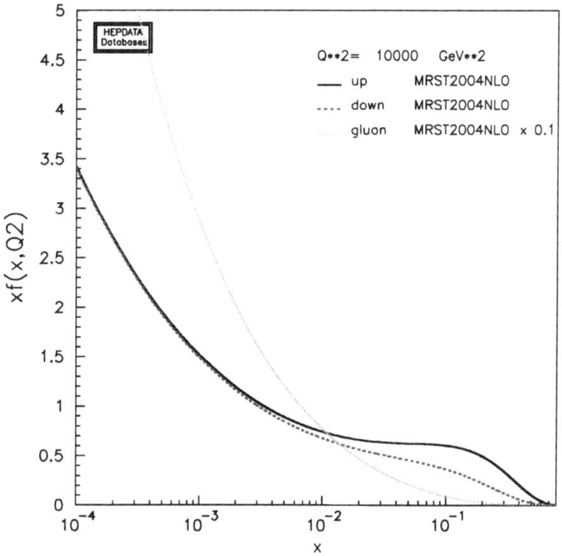

Fig. 2.2 Proton parton distribution functions multiplied with x. The gluon distribution is scaled with a factor 1/10 [11]

2.3 Hard Scattering Processes in Hadron Collisions

When two hadrons collide at high energy, most of the collisions involve only soft interactions of the constituent quarks and gluons. Such interactions cannot be treated using perturbative QCD, because α_s is large when the momentum transfer is small. In some collisions however, two quarks or gluons will exchange a large momentum. In those cases, the elementary interaction takes place very rapidly compared to the internal time scale of the hadron wavefunctions, so the lowest order(s) QCD prediction should accurately describe the process.

The cross section for such a process can be written as a factorized product of short and long distance processes:

$$\sigma(P_1, P_2) = \sum_{i,j} \int dx_1 dx_2 f_i(x_1, \mu_F^2) f_j(x_2, \mu_F^2) \hat{\sigma}_{i,j}(\mu_R^2, \mu_F^2), \tag{2.19}$$

where P_1 and P_2 denote the momenta of the incoming hadrons. Figure 2.3 shows this schematically. The momenta of the partons that participate in the hard interaction are $p_1 = x_1 P_1$ and $p_2 = x_2 P_2$. The functions $f_i(x_1, \mu_F^2)$ and $f_j(x_2, \mu_F^2)$ are the usual QCD quark or gluon PDFs, defined at a factorization scale μ_F, which take into account the long-distance effects. It is in this sense that μ_F can be thought of as the scale which separates long- and short-distance physics.

The short-distance cross section for the scattering of partons of types i and j is denoted by $\hat{\sigma}_{i,j}$. Since the coupling is small at high energy, $\hat{\sigma}_{i,j}$ can be calculated as a perturbation series in α_s.

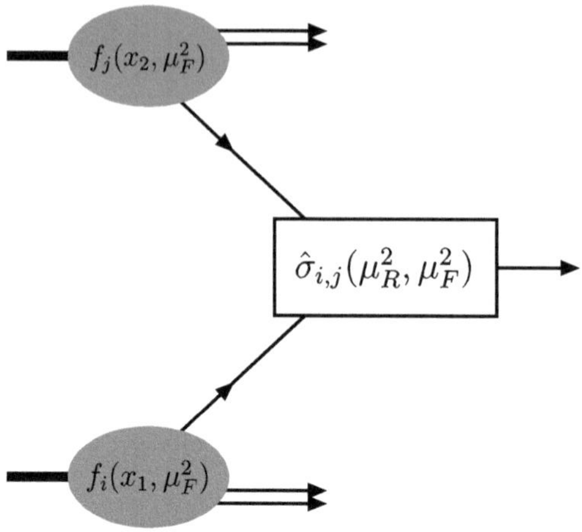

Fig. 2.3 The parton model description of a hard scattering process

At leading order, $\hat{\sigma}_{i,j}$ is identical to the normal parton scattering cross section and the dependence on μ_F disappears, but at higher order, long-distance parts in the parton cross section need to be removed and factored into the parton distribution functions.

Note that if calculated to all orders, the cross section should be independent of the factorization and renormalization scales:

$$\frac{\partial \sigma}{\partial \mu_F} = \frac{\partial \sigma}{\partial \mu_R} = 0 \tag{2.20}$$

In practice, one is restricted to calculations at low orders, for which the residual dependence on μ_F and μ_R can be appreciable.

Equation 2.19 is a prediction of the cross section with partons in the outgoing state. Experiments however, measure hadrons and not partons due to confinement. The non-perturbative process that transforms partons into hadrons is called hadronization and this will be discussed in Sect. 2.5. But first we will discuss parton showers in the next section.

2.4 Parton Branching

As discussed in Sect. 2.3, the hard collision between two hadrons, can be understood as the collision between two partons. The first terms in the perturbative QCD expansion, usually suffice to describe successfully the hard interaction between these two partons, because the scale of this process is large.

However, in some regions of the phase space, higher order terms are enhanced and cannot be neglected. For example, we have seen in Sect. 2.2 about the parton model that when a quark emits a gluon, perturbation theory fails to describe the process in the collinear region.

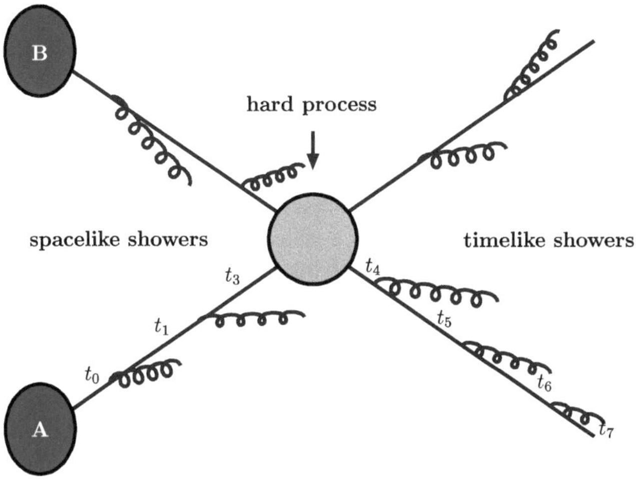

Fig. 2.4 Schematic illustration of the hard scattering process and the softer showers. For initial state branchings, t is increasing towards the hard scattering by means of successive small-angle emissions ($t_0 < t_1 < t_3$). The opposite is true for final state branching, where t is decreasing after every branching ($t_4 > t_5 > t_6 > t_7$)

Enhanced higher-order terms occur in processes where a soft gluon is emitted or when a gluon or light quark splits into two almost collinear partons. *Parton branching* is the common name for these soft and collinear configurations.

In collision processes, parton branching typically happens for the ingoing and outgoing quarks and gluons of the hard interaction. The incoming quark, initially with low virtual mass-squared and carrying a fraction x of the hadron's momentum, moves to more virtual masses and lower momentum fractions by successive small-angle emissions, and finally undergoes the hard scattering which happens at a scale Q. After the collision, the outgoing parton of the hard scattering process has initially a large positive mass-squared, which then gradually decreases by consecutive parton emissions.

Figure 2.4 shows schematically a hard hadron collision. Two hadrons (A and B) are coming in and one incoming parton in each hadron gets selected, and undergoes a hard scattering, resulting in outgoing partons. The hard scattering of the incoming partons which happens at a scale Q, can be calculated using perturbative QCD. But all incoming and outgoing partons undergo branchings as well, giving rise to the so called parton showers (and to scale dependent PDFs). A lower order perturbative calculation fails to describe the shower behavior, but perturbative QCD calculations become too complicated at higher orders to be of practical use. We will show that an approximate perturbative treatment of QCD to all orders is adequate at describing the branching physics.

A distinction needs to be made between partons that are incoming lines in the Feynman diagram describing the hard interaction, and partons that are outgoing lines. An incoming parton has a negative (virtual) mass-squared. Therefore its branching process is called spacelike, giving rise to initial state showers. The opposite is true for outgoing branching partons. These partons have a positive mass-squared and their branching is said to be timelike. They give rise to final state showers.

A branching can be seen as a $a \to bc$ process, where A is called the mother and b and c the daughters. Each daughter is free to branch as well, so that a shower-like structure can evolve.

For a timelike branching, we assume that the mass of the mother is much higher than the masses of the daughters. For a spacelike branching, we assume that the daughter that will finally take part in the hard interaction has a much larger virtuality than the other partons.

In the approximation of small angle scattering, the branching kinematics can be described by two variables, z and t. We define z as the fraction of energy carried by daughter b: $z = E_b/E_a = 1 - E_c/E_a$. The variable t can have different interpretations, but always has the dimensions of squared mass. Here we will define t as the mass squared of the mother ($t \equiv p_a^2$) for timelike branching and as the absolute value of the mass squared of the daughter ($t \equiv |p_b^2|$) for spacelike branching.

In terms of z and t, the differential probability that one branching occurs is given by:

$$d\mathcal{P}_a = \sum_{b,c} \frac{\alpha_s}{2\pi} P_{a\to bc}(z) \frac{dt}{t} dz, \tag{2.21}$$

where the sum runs over all branchings the parton is allowed to make. The functions $P_{a\to bc}(z)$ are the so called splitting kernels. They are written as a perturbation series and, at lowest order, can be interpreted as the probability of finding a parton of type b in a parton of type a with a momentum fraction z. For example, for the splitting of a gluon into a quark antiquark pair, we have at lowest order that $P_{g\to q\bar{q}}(z) \propto (z^2 + (1-z)^2)$. We integrate Eq. 2.21 over z in order to get the branching probability for a certain t value:

$$\mathcal{I}_{a\to bc}(t) = \int_{z^-(t)}^{z^+(t)} dz \frac{\alpha_s}{2\pi} P_{a\to bc}(z), \tag{2.22}$$

where we have considered one type of branching only. In principle z can vary between 0 and 1, but because most splitting kernels suffer from infrared singularities at $z = \{0, 1\}$, we need to introduce an explicit cut-off. Physically, this can be understood by saying that branchings close to the integration limits are unresolvable; they involve the emission of an undetectably soft parton. Alternatively, the plus prescription of the splitting function can be used instead of $z^-(t)$ and $z^+(t)$ [7–9].

The naïve probability that a branching occurs in the range $[t, t + dt]$, is given by $\sum_{b,c} \mathcal{I}_{a\to bc}(t)dt/t$, and thus the probability of no emission is $1 - \sum_{b,c} \mathcal{I}_{a\to bc}(t)dt/t$. This is however, not correct when we consider multiple branchings. Note that from Heisenberg's principle, t fills the function of a kind of inverse time squared for the shower evolution; t is constrained to be gradually decreasing away from the hard scattering in final state showers, and to be gradually increasing towards the hard scattering in initial state showers.

This means that the probability for branching at a time t needs to take into account the probability that the parton has not branched at earlier times $t_0 < t$. The probability that a branching did not occur between t_0 and t, is given by the Sudakov form factor [13]:

$$\mathcal{P}_{\text{no-branching}}(t_0, t) = \exp\left\{ -\int_{t_0}^{t} \frac{\mathrm{d}t'}{t'} \sum_{b,c} \mathcal{I}_{a\to bc}(t') \right\} = S_a(t), \tag{2.23}$$

giving rise to the actual probability that a branching occurs at time t:

$$\begin{aligned}\frac{\mathrm{d}\mathcal{P}_a}{\mathrm{d}t} &= -\frac{\mathrm{d}\mathcal{P}_{\text{no-branching}}(t_0, t)}{\mathrm{d}t} = \Big(\frac{1}{t}\sum_{b,c}\mathcal{I}_{a\to bc}(t)\Big) S_a(t) \\ &= \Big(\frac{1}{t}\sum_{b,c}\int_{z^-(t)}^{z^+(t)} \mathrm{d}z \frac{\alpha_s}{2\pi} P_{a\to bc}\Big)\exp\left\{-\int_{t_0}^{t}\frac{dt'}{t'}\sum_{b,c}\int_{z^-(t')}^{z^+(t')} \mathrm{d}z\frac{\alpha_s}{2\pi}P_{a\to bc}\right\}\end{aligned} \tag{2.24}$$

The first term in the right hand side of the above equation is the naïve branching probability. The other term is needed to deal with the fact that partons that have already branched can no longer branch. This is similar to the radioactive decay.

Equation 2.24 can be used to simulate jet production, and therefore forms the basis for parton showers implemented in many Monte Carlo event generators [14].

Because inside the hadron, sea quarks and gluons undergo the same branchings as described in this section, the evolution of PDFs can be described with the same techniques [1]. These are the DGLAP equations, which were shown in Sect. 2.2 (see Eq. 2.18).

The DGLAP equations are not applicable in all regions of phase space. As a matter of fact, it turns out that when $\ln(t/\Lambda^2) \ll \ln(1/x)$, i.e. for small values of x, not all leading terms are included; important contributions in terms of $\ln(1/x)$ are neglected. The resummation of terms proportional to $\alpha_s \ln(1/x)$ to all orders, retaining the full t dependence and not just the leading $\ln(t)$ is accomplished by the Balitsky–Fadin–Kuraev–Lipatov (BFKL) [15, 16] equation.

2.5 Hadronization

Due to color confinement, quarks and gluons cannot propagate freely over macroscopic distances. When two quarks are close together, the strong force between them is relatively weak (asymptotic freedom), but when they move farther apart, the force becomes much stronger (confinement). The potential between the quarks increases linearly with their mutual separation, and at some distance, it becomes much easier to create a new quark-antiquark pair than to keep pulling against the ever-increasing potential. This process is repetitive and the newly created quarks and antiquarks will combine themselves into hadrons.

In a collision experiment, all outgoing partons will therefore undergo parton showering and transform themselves into hadrons, forming jets, i.e. sprays of hadrons, which are then experimentally detected. The process is called *hadronization*.

Hadronization cannot be calculated in perturbative QCD, because it happens in a region where α_s is too strong. But still, jets are very useful for our understanding of QCD. The reason is that by the uncertainty principle, a hard interaction at a typically large scale Q occurs at a distance scale of the order of $1/Q$, while the subsequent hadronization processes occur at a much later time scale characterized by $1/\Lambda$, where Λ is the scale at which the strong coupling becomes strong. The interactions that change quarks and gluons into hadrons, certainly modify the outgoing state, but they occur too late to modify the original probability for the event to happen, which can therefore be calculated in perturbation theory. Each hadron appears in the final state roughly in the same direction as the quark or gluon it originated from. The cross section for a single hadron is therefore closely related to the underlying partonic direction, and for a good jet finding algorithm, the extension to jet cross sections can be made. We will talk about jets in detail in later chapters.

Popular models describing hadronization are the Lund string model [17] and the cluster model [18]. In all models, color singlet structures are formed out of color connected partons, and are decayed into hadrons preserving energy and momentum.

2.6 Monte Carlo Event Generators

As already mentioned in the introductory chapter, particle collision experiments are of high importance for testing theories. In order to be able to interpret scattering experiments in terms of an underlying theory, a comparison between events simulated according to that specific theory and data is needed. Since nature is fundamentally probabilistic, the generated events need to exhibit the same statistical fluctuations. Pseudo-randomness can be computed using suited Monte Carlo techniques.

The generation of an event is done using a factorized approach, and the major steps are:

1. the hard scattering process
2. initial and final state radiation (i.e. parton showers)
3. hadronization and beam remnants
4. multiple interactions

The first three steps were discussed in this chapter, but more generator-specific information can be found in Ref. [14].

Besides a hard scattering, additional interactions between partons occur in the event, which are called multiple interactions and cannot be neglected.

A beam remnant is what remains of the incoming beam after one of its partons has initiated the hard scattering. Because the beam remnants are no longer color neutral, they need to be included into the calculation.

Due to its high complexity, the hard scattering is usually calculated at leading order. Programs with higher order scatterings exist, but these programs do not include the other steps of the event generation (i.e. they are not *complete*).

The work in this thesis is done using four generators:

- PYTHIA [14, 19]
- NLOJET++ [20]
- JETRAD [21]
- GravADD [22]

PYTHIA is a complete, multi-purpose event generator with leading-order matrix elements. Within many experimental collaborations, this program has become the standard for providing event properties in a wide range of reactions, within and beyond the Standard Model, with emphasis on those that include strong interactions, directly or indirectly, and therefore multihadronic final states. While the first releases were coded in Fortran [14], more current releases have been written in C++ [19] .

NLOJET++ and JETRAD use a next-to-leading order (NLO) description of the hard scattering, but parton showers, hadronization, beam remnants and multiple interactions are not implemented. NLO Monte Carlo techniques will be the topic of Chap. 3 .

GravADD is a complete generator for black holes and gravitational scattering in large extra dimensions, in addition to standard QCD processes. See Chap. 5 for a detailed description.

References

1. R.W. Ellis, W.J. Stirling, B.R. Webber, *QCD and Collider Physics*. Cambridge Monographs on Particle Physics Nuclear Physics and Cosmology (Cambridge University Press, Cambridge, 1996)
2. M.E. Peskin, D.V. Schroeder, *An Introduction to Quantum Field Theory* (Westview Boulder, CO, 1995)
3. C. Amsler et al., Review of particle physics. Phys. Lett. **B667**, 1 (2008)
4. S. Bethke, α_s at Zinnowitz 2004. Nuclear Physics B- Proceedings Supplements, **135** 345–352 hep-ex/0407021 (2004)
5. J.D. Bjorken, Asymptotic sum rules at infinite momentum. Phys. Rev. **179**, 1547–1553 (1969)
6. R.P. Feynman, *Photon-Hadron Interactions* (WA Benjamin Reading Mass, New York, 1972)
7. G. Altarelli, G. Parisi, Asymptotic freedom in parton language. Nucl. Phys. **B126**, 298 (1977)
8. Y.L. Dokshitzer, Sov. Phys. JETP **46**, 641 (1977)
9. V.N. Gribov, L.N. Lipatov, Yad. Fiz. **15**, 781 (1972)
10. P.M. Nadolsky, H.L. Lai, Q.-H. Cao, J. Huston, J. Pumplin, D. Stump, W.K. Tung, C.P. Yuan, Implications of CTEQ global analysis for collider observables. Phys. Rev. **D78**, 013004 (2008)
11. A.D. Martin, R.G. Roberts, W.J. Stirling, R.S. Thorne, Physical gluons and high-ET jets. Phys. Lett. **B604**, 61–68 (2004) (hepph0410230)

12. A.D. Martin, W.J. Stirling, R.S. Thorne, G. Watt, Parton distributions for the LHC. EPJ C **63**, 189–285 (2009)
13. V.V. Sudakov, Vertex parts at very high-energies in quantum electrodynamics. Sov. Phys. JETP **3**, 65–71 (1956)
14. T. Sjöstrand, S. Mrenna, P. Skands, PYTHIA 6.4 physics and manual. JHEP **0605**, 026 (2006) (hep-ph0603175)
15. E.A. Kuraev, L.N. Lipatov, V.S. Fadin, The pomeranchuk singularity in nonabelian gauge theories. Sov. Phys. JETP **45**, 199–204 (1977)
16. Y.Y. Balitsky, L.N. Lipatov, The pomeranchuk singularity in quantum chromodynamics. Sov. J. Nucl. Phys. **28**, 822–829 (1978)
17. B. Andersson, G. Gustafson, G. Ingelman, T. Sjostrand, Parton fragmentation and string dynamics. Phys. Rept. **97**, 31–145 (1983)
18. R.D. Field, S. Wolfram, A QCD model for e^+e^- annihilation. Nucl. Phys. B **213**, 65–84 (1983)
19. T. Sjöstrand, S. Mrenna, P. Skands, A brief introduction to PYTHIA 8.1. JHEP **05**, 026 (2007) (hep-ph07103820)
20. Z. Nagy, Next-to-leading order calculation of three-jet observables in hadron–hadron collision. Phys. Rev. **D68**, 094002 (2003) (hep-ph0307268)
21. W.T. Giele, E.W.N. Glover, D.A. Kosower, Higher order corrections to jet cross sections in hadron colliders. Nucl. Phys. B **B403**, 633 (1993) (hep-ph9302225)
22. L. Lönnblad, M. Sjödahl, Classical and non-classical ADD-phenomenology with high-ET jet observables at collider experiments. JHEP **0610**, 088 (2006) (hep-ph/0608210)

Chapter 3
NLO Monte Carlo Techniques

3.1 Introduction

Although LO calculations generally describe broad features of a particular process and provide the first estimate of its cross section, in many cases this approximation is insufficient. The inherent uncertainty in a lowest-order calculation derives from its dependence on the unphysical renormalization and factorization scales, which is often large. In addition, some processes may contain large logarithms that need to be resummed, or extra partonic processes may contribute only when going beyond the first approximation.

Thus, in order to compare data with predictions that have smaller theoretical uncertainties, next-to-leading order calculations are a must. This chapter will present the tools for next-to-leading order (NLO) calculations.

For simplicity, we will start with discussing processes with no initial state hadrons (e.g. in e^+e^- annihilation). The cross section can be written as an expansion in the strong coupling constant:

$$\sigma = \sigma_0 + \sigma_1\alpha_s(Q) + \sigma_2\alpha_s^2(Q) + \mathcal{O}\left(\alpha_s^3(Q)\right) \tag{3.1}$$

Using Eq. 2.11 to express this as function of $\alpha_s(\mu)$, Eq. 3.1 becomes

$$\sigma = \sigma_0 + \sigma_1\alpha_s(\mu) + \left(\sigma_2 - \sigma_1\frac{\beta_0}{2\pi}\ln(Q^2/\mu^2)\right)\alpha_s^2(Q) + \mathcal{O}\left(\alpha_s^3(Q)\right) \tag{3.2}$$

Since μ is arbitrary and therefore $\alpha_s(\mu)$ can be given any value, a leading-order QCD calculation of the cross section (i.e. the first two terms in Eq. 3.2) predicts only the order of magnitude. To get some control over the scale dependence, at least a next-to-leading order calculation (i.e.including the third term in Eq. 3.2) is required.

Let us assume that we want to compute the next-to-leading order m-jet cross section, i.e. the cross section for m jets in the final states obtained from running a jet algorithm on the final-state partons. Up to NLO, the cross section σ can be written as:

N. Boelaert, *Dijet Angular Distributions in Proton–Proton Collisions*, Springer Theses, DOI: 10.1007/978-3-642-24597-8_3, © Springer-Verlag Berlin Heidelberg 2012

$$\sigma = \sigma^{LO} + \sigma^{NLO} \tag{3.3}$$

The LO cross section involves m partons in the final state:

$$\sigma^{LO} = \int_m \mathrm{d}\sigma^B, \tag{3.4}$$

where $d\sigma^B$ stands for the Born approximation.

At NLO, the m-jet cross section receives contributions from virtual corrections to the m-parton final state ($d\sigma^V$), and from real corrections coming from the $(m+1)$-parton final state ($d\sigma^R$):

$$\sigma^{NLO} \equiv \int \mathrm{d}\sigma^{NLO} = \int_{(m+1)} \mathrm{d}\sigma^R + \int_m \mathrm{d}\sigma^V \tag{3.5}$$

Both contributions are separately divergent in $d = 4$ dimensions, though their sum is finite.

Two conceptually different techniques have been developed for dealing with these divergences, namely the *phase space slicing* technique [1] and the *subtraction* scheme [2]. The phase space slicing method is based on approximating the matrix elements and the phase space integration in boundary regions of phase space so that the integration may be carried out analytically. The subtraction method is based on adding and subtracting counterterms designed to approximate the real emission amplitudes in the phase space boundary regions on the one hand, and to be integrable with respect to the momentum of an unresolved parton on the other hand. The most recent implementation is the dipole subtraction method [3]. We will discuss both techniques in the next two sections, and we will make a brief comparison in Sect. 3.4.

The same principles hold for processes with initial state hadrons, but in these processes, additional soft and collinear singularities in the initial state are absorbed in the PDFs and give rise to extra counterterms.

Note that the virtual contribution in Eq. 3.5 suffers from ultraviolet poles as well, but we assume that they have been removed by carrying out a renormalization procedure.

3.2 The Dipole Subtraction Method

3.2.1 General Method

The general idea behind this method is to rewrite Eq. 3.5 in the following way:

$$\sigma^{NLO} = \int_{(m+1)} \left[\mathrm{d}\sigma^R - \mathrm{d}\sigma^A\right] + \int_{(m+1)} \mathrm{d}\sigma^A + \int_m \mathrm{d}\sigma^V, \tag{3.6}$$

where $d\sigma^A$ functions as a local counterterm for $d\sigma^R$; it is an approximation of $d\sigma^R$ in the sense that it has the same pointwise singular behavior as $d\sigma^R$ itself.

The first term of the right-hand side of Eq. 3.6 can be integrated numerically in four dimensions. All singularities are associated with the last two terms. Using dimensional regularization in $d = 4 - 2\epsilon$ dimensions, the virtual divergences are replaced by poles in ϵ. But these poles can be combined with the ones resulting from the analytical integration of $d\sigma^A$ over the one-parton phase subspace, canceling all divergences. This cancelation is however, only guaranteed for cross sections of so-called *jet observables*, i.e. hadronic observables that are defined in such a way that their actual value is independent of the number of soft and collinear hadrons (partons) produced in the final state. In particular, this means that the jet observable has to be the same in a given m-parton configuration and in all $(m + 1)$-parton configurations that are kinematically degenerate with it (i.e. that are obtained from the m-parton configuration by adding a soft parton or replacing a parton with a pair of collinear partons carrying the same total momentum). We will discuss this issue more in Chap. 7, but this means that the collinear and infrared safety of a jet algorithm is a must.

After this cancelation, the limit $\epsilon \to 0$ can be carried out without problems and the remaining integration over the m-parton phase space can be calculated numerically. Schematically:

$$\sigma^{NLO} = \int_{(m+1)} \left[\left(d\sigma^R \right)_{\epsilon=0} - \left(d\sigma^A \right)_{\epsilon=0} \right] + \int_m \left[d\sigma^V + \int_1 d\sigma^A \right]_{\epsilon=0} \tag{3.7}$$

The challenging task of the above subtraction scheme is to create a method to construct the actual form of $d\sigma^A$. Using the physical knowledge of how $(m + 1)$-parton matrix elements behave in the soft and collinear limits, so-called universal dipole factors can be constructed which allow for $d\sigma^A$ to be rewritten in a factorized form:

$$d\sigma^A = \sum_{\text{dipoles}} d\sigma^B \otimes dV_{\text{dipole}} \tag{3.8}$$

The notation in Eq. 3.8 is symbolic. Here $d\sigma^B$ denotes an appropriate color and spin projection of the Born-level exclusive cross section. The symbol $\otimes$ stands for properly defined phase space convolutions and sums over color and spin indices. The dipole factors dV_{dipole}, which match the singular behavior of $d\sigma^R$, are universal, i.e. completely independent of the details of the process and they can be computed once for all. The dependence on the jet observable is completely embodied by the factor $d\sigma^B$.

There are several dipole terms on the right-hand side of Eq. 3.8, each of them corresponding to a different kinematic configuration of $(m + 1)$ partons. Each configuration can be thought of as obtained by an effective two-step process; an m-parton configuration is first produced and then a dipole of two massless partons decays into three partons. It is this two-step pseudo-process that leads to the factorized structure on the right-hand side of Eq. 3.8. This means that whenever the $(m+1)$-parton state in $\mathrm{d}\sigma^R$ approaches a soft and/or collinear region, there is a corresponding dipole factor in $\mathrm{d}\sigma^A$ that approaches the same region with exactly the same probability as in $\mathrm{d}\sigma^R$. In this manner $\mathrm{d}\sigma^A$ acts as a local counterterm for $\mathrm{d}\sigma^R$.

The product structure in Eq. 3.8 allows for a factorizable mapping from the $(m+1)$-parton phase space to an m-parton subspace times a single-parton phase space:

$$\int_{(m+1)} \mathrm{d}\sigma^A = \sum_{\text{dipoles}} \int_m \mathrm{d}\sigma^B \otimes \int_1 \mathrm{d}V_{\text{dipole}} = \int_m \left[\mathrm{d}\sigma^B \otimes \boldsymbol{I}\right], \tag{3.9}$$

where all poles are contained in the universal factor $\boldsymbol{I}$:

$$\boldsymbol{I} = \sum_{\text{dipoles}} \int_1 \mathrm{dV}_{\text{dipole}} \tag{3.10}$$

The final result can be written as:

$$\begin{aligned}\sigma^{NLO} &= \sigma^{NLO\{(m+1)\}} + \sigma^{NLO\{m\}} \\ &= \int_{(m+1)} \left[\left(d\sigma^R\right)_{\epsilon=0} - \left(\sum_{\text{dipoles}} \mathrm{d}\sigma^B \otimes \mathrm{d}V_{\text{dipole}}\right)_{\epsilon=0}\right] \\ &\quad + \int_m \left[\mathrm{d}\sigma^V + \mathrm{d}\sigma^B \otimes \boldsymbol{I}\right]_{\epsilon=0}\end{aligned} \tag{3.11}$$

The above subtraction scheme however, only holds for processes with no initial-state hadrons. But with a few modifications, it can also be used for hadron–hadron collisions. For hadron–hadron processes, the cross section can be regarded as a convolution of the partonic (short-distance) cross section with non-perturbative parton density functions (Eq. 2.19). The partonic cross section can be written in a similar way as done in Eq. 3.5, but a collinear counterterm $\int_1 d\sigma^C$ needs to be added in order to account for the factorization scale dependency of the parton densities:

$$\sigma^{NLO} \equiv \int \mathrm{d}\sigma^{NLO} = \int_{(m+1)} \mathrm{d}\sigma^R + \int_m \left[\mathrm{d}\sigma^V + \mathrm{d}\sigma^C\right] \tag{3.12}$$

We can apply the subtraction method described above (Eq. 3.7) to evaluate this cross section:

$$\sigma^{NLO} = \int_{(m+1)} \left[\left(\mathrm{d}\sigma^R\right)_{\epsilon=0} - \left(\mathrm{d}\sigma^A\right)_{\epsilon=0}\right] + \int_m \left[\mathrm{d}\sigma^V + \mathrm{d}\sigma^C + \int_1 \mathrm{d}\sigma^A\right]_{\epsilon=0} \tag{3.13}$$

However, a few modifications in the construction of $d\sigma^A$ need to be made. We will discuss them briefly.

In hadron–hadron collisions, an extra singularity for the real cross section $d\sigma^R$ occurs when one of the $(m + 1)$ final-state partons becomes collinear to a parton in the initial state. Furthermore, because of the well defined momenta of the initial-state partons, the phase space integration has to be performed in the presence of additional kinematics constraints. In order to make sure that the counterterm $d\sigma^A$ is also canceling these later singularities and that the integral $\int_1 d\sigma^A$ is still computable analytically, even in the presence of the additional phase space constraints, Eq. 3.8 needs to be modified:

$$d\sigma^A = \sum_{\text{dipoles}} d\sigma^B \otimes \left[dV_{\text{dipole}} + dV'_{\text{dipole}}\right] \tag{3.14}$$

The additional dipole terms dV'_{dipole} match the singularities of $d\sigma^R$ coming from the region collinear to the momenta of the initial partons. They are also integrable analytically over the one-parton subspace.

Using the above expression for $d\sigma^A$, the dipole subtraction scheme in Eq. 3.13 can now be rewritten as:

$$\begin{aligned}
\sigma^{NLO}(p) &= \sigma^{NLO\{(m+1)\}}(p) + \sigma^{NLO\{m\}}(p) + \int_0^1 dx \hat{\sigma}^{NLO\{m\}}(x; xp) \\
&= \int_{(m+1)} \left[\left(d\sigma^R(p)\right)_{\epsilon=0} - \left(\sum_{\text{dipoles}} d\sigma^B(p) \otimes \left(dV_{\text{dipole}} + dV'_{\text{dipole}}\right)\right)_{\epsilon=0}\right] \\
&\quad + \int_m \left[d\sigma^V(p) + d\sigma^B(p) \otimes \boldsymbol{I}\right]_{\epsilon=0} \\
&\quad + \int_0^1 dx \int_m \left[d\sigma^B(xp) \otimes (\boldsymbol{P} + \boldsymbol{K} + \boldsymbol{H})(x)\right]_{\epsilon=0}
\end{aligned} \tag{3.15}$$

The momentum p denotes the dependence on the momenta of the incoming partons and x is the longitudinal momentum fraction. The contributions $\sigma^{NLO\{(m+1)\}}(p)$ and $\sigma^{NLO\{m\}}(p)$ are completely analogous to those in Eq. 3.11. The last term is a finite remainder that is left after factorization of initial-state and final-state collinear singularities into the non-perturbative parton density functions. The functions $\boldsymbol{P}$, $\boldsymbol{K}$ and $\boldsymbol{H}$ are similar to $\boldsymbol{I}$, that is, they are universal—independent of the detail of the scattering process and of the jet observable—and depend on the number of initial-state partons only.

3.2.2 NLOJET++

NLOJET++ [4] is a multipurpose C++ program for calculating jet cross sections in e^+e^- annihilations, DIS and hadron-hadron collisions: $e^+e^- \rightarrow$ 4 jets, $ep \rightarrow (\leq 3+1)$ jets, $p\bar{p} \rightarrow \leq 3$ jets. Its core library is based on the dipole subtraction method discussed in the previous section, but with a modification that was implemented for computational reasons. NLOJET++ uses a cut dipole phase space parameter $\alpha \in [0, 1]$ to control the volume of the dipole phase space, with the original dipole subtraction scheme obtained for $\alpha = 1$. The NLO corrections are independent of the value of α, but $\alpha < 1$ is favored because of computational reasons. This is because a smaller value of α speeds up the evaluation of the subtraction terms and increases the stability of the program because it reduces the chance of missed binning. Missed binning happens when a very large positive weight from the real part and the corresponding weight from the subtraction term are filled into different histogram bins. The introduction of this dipole phase space parameter requires a redefinition of the functions $\boldsymbol{K}$ and $\boldsymbol{I}$. We refer to the reference for the exact implementation.

3.3 The Phase Space Slicing Technique

3.3.1 General Method

The core of this method is to divide the $(m + 1)$-phase space into regions where $(m + 1)$ partons are "resolved" and regions where only m-partons are "resolved". The invariant mass s_{ij} between two partons is used to make the distinction; when s_{ij} is smaller than a resolution factor s_{min}, only one parton is resolved, but both partons are resolved in the other case. The reason for doing this is that all divergences occurring in the real $(m+1)$-parton final state are associated with the regions where only m-partons are resolved. In this soft/collinear region, both matrix element and phase space can be approximated, and the integration over the unresolved degrees of freedom can be done analytically, so that these real divergences can be canceled directly against the virtual corrections of the m-parton cross section.

Rather than using the full matrix element, the ordered amplitudes associated with each color structure are used, and the full squared amplitude is obtained by summing the squared ordered amplitudes over all permutations of the identical particles in the final state, neglecting subleading terms in the number of colors. Symbolically:

$$|\mathcal{M}|^2 \sim \sum_{Perm.} |A|^2 . \tag{3.16}$$

For the calculation of the real corrections, the $(m+1)$-phase space is decomposed into two regions; the resolved region, R, and the unresolved region, U, with the partitioning depending on the resolution factor s_{min}. This way, the real emission scattering amplitude is split up into a resolved and an unresolved part.

The calculation of the resolved part is performed using standard Monte Carlo techniques, and the result depends logarithmically on $s_{\min}$.

The fact that both the phase space and the matrix element exhibit a factorization of the soft and collinear parts in the unresolved phase space regions, meaning that the integration of the $(m+1)$-amplitude over the singular (or unresolved) parts of phase space can be done analytically in $d = 4 - 2\epsilon$ dimensions, without knowing the $(m+1)$-scattering amplitude explicitly, can be used to calculate the unresolved part:

$$\int \mathrm{d}\,\mathrm{PS}_{\mathrm{soft/collinear}}\, |A_{n+1}|^2 \rightarrow R(s_{\min})\, |A_n|^2 \tag{3.17}$$

The result of this integration is proportional to the tree level amplitude, and contains double and/or single poles in ϵ and accompanying double and/or single logarithms in $s_{\min}$. For sufficiently small values of $s_{\min}$, the $s_{\min}$ terms in the U region compensate the $s_{\min}$ logarithms in the R region.

Furthermore, soft and collinear final state poles in ϵ in the U region are canceled against the poles that show up in the virtual corrections. The virtual corrections to the squared matrix elements have the generic form:

$$|\mathcal{M}|_V^2 = \sum_{Perm.} V\, |A_n|^2 + \mathcal{F}, \tag{3.18}$$

where V is the singular part proportional to the three level ordered amplitude and $\mathcal{F}$ is the remaining finite contribution. Combining Eq. 3.18 with Eq. 3.17 results in the finite (F) next-to-leading order squared matrix elements:

$$|\mathcal{M}|_F^2 \sim \sum_{Perm.} \left([1 + \mathcal{K}]\, |A_n|^2 + \mathcal{F} \right) \tag{3.19}$$

The combination of the phase space factor and the virtual factor, $\mathcal{K} = R + V$, is finite (this is guaranteed by the Kinoshita–Lee–Nauenberg theorem [5–7]) and therefore the squaring and summation over all polarizations can be performed in four dimensions. While $\mathcal{K}$ is universal, the virtual correction $\mathcal{F}$ is process dependent.

As for the collinear singularities in the initial state, they are absorbed into the bare PDFs, leaving a finite remainder which is written in terms of modified PDFs.

Provided that $s_{\min}$ is small enough so that the soft and collinear approximations are valid and that $s_{\min}$ is not so small that the calculation becomes numerically unstable, the cross section should be finite and independent of the unphysical $s_{\min}$. A reasonable value to choose for processes at LHC energies, is $s_{\min} = 10\,\mathrm{GeV}^2$.

3.3.2 JETRAD

JETRAD [1] is an NLO Monte Carlo generator for 1-jet and 2-jet production at hadron colliders, it is a direct implementation of the method described in the previous section. The program uses the VEGAS-algorithm [8] to perform the integration.

3.4 Comparison of the Subtraction Method and the Phase Space Slicing Technique

The difference between the methods was summarized in Ref. [9]. Schematically, the $(m+1)$-parton contribution to the m-jet cross section is given by

$$\begin{aligned} \mathrm{d},\sigma_{(m+1)} &= |\mathcal{M}_{(m+1)}|^2 \times J_{(m+1)} \mathrm{d}\,\mathrm{PS}_{(m+1)} \\ &= \left(|\mathcal{M}_{(m+1)}|^2 \times (1-\theta_s) + |\mathcal{M}_{(m+1)}|^2 \times \theta_s\right) \times J_{(m+1)} d,\mathrm{PS}_{(m+1)} \\ &= |\mathcal{M}_{(m+1)}|^2 \times (1-\theta_s) J_{(m+1)} d\mathrm{PS}_{(m+1)} \\ &\quad + \theta_s \times (T_1(\theta_s) + T_2(\theta_s) + T_3(\theta_s)), \end{aligned} \tag{3.20}$$

where the $(m+1)$-parton differential cross section $\mathrm{d}\sigma_{(m+1)}$ is given by the matrix element squared, $|\mathcal{M}_{(m+1)}|^2$, $J_{(m+1)}$ which is a term including phase space constraints from the jet algorithm and cuts, and $\mathrm{d}\,\mathrm{PS}_{(m+1)}$ which is the $(m+1)$-parton phase space. The soft/collinear unresolved part of phase space is separated off using the resolution criterion embodied in the quantity θ_s, which takes the value $0_s = 1$ in the unresolved phase space region and $\theta_s = 0$ otherwise.

T_1 is given by

$$\begin{aligned} T_1(\theta_s) &= S|\mathcal{M}_m|^2 \times J_m \mathrm{d}\,\mathrm{PS}_{\mathrm{soft}} d\mathrm{PS}_m \\ &= R(\theta_s)|\mathcal{M}_m|^2 \times J_m d\mathrm{PS}_m, \end{aligned} \tag{3.21}$$

and represents the integral of the approximate matrix element $|\mathcal{M}_{(m+1)}|^2 \to S|\mathcal{M}_m|^2$ over the approximate phase space $\mathrm{d}\,\mathrm{PS}_{(m+1)} \to \mathrm{d}\,\mathrm{PS}_{\mathrm{soft}} \mathrm{d}\,\mathrm{PS}_m$. The resolution factor $R(\theta_s)$ is independent of the hard scattering and can be calculated analytically for a wide range of multiparton processes.

T_2 is given by

$$T_2(\theta_s) = \left(|\mathcal{M}_{(m+1)}|^2 - S|\mathcal{M}_m|^2\right) \times J_{(m+1)} \mathrm{d}\,\mathrm{PS}_{(m+1)}, \tag{3.22}$$

and represents the integral over the true unresolved phase space of the difference between the true matrix element and the approximate matrix element.

Finally, T_3 is given by

$$T_3(\theta_s) = S|\mathcal{M}_m|^2 \left(J_{(m+1)} \mathrm{d}\,\mathrm{PS}_{(m+1)} - J_m \mathrm{d}\,\mathrm{PS}_m \mathrm{d}\,\mathrm{PS}_{soft}\right), \tag{3.23}$$

and represents the difference between the integrals of the approximate matrix element over the true unresolved phase space and the approximate unresolved phase space. Note that T_1 contains the soft and collinear divergences needed to cancel the singularities of the virtual term, while T_2 and T_3 vanish as θ_s is taken to zero.

The phase space slicing method keeps T_1, but sets $T_2 = T_3 = 0$, while the subtraction method keeps both T_1 and T_2, but sets $T_3 = 0$. An exact method, in which both the correction factors for the phase space and matrix elements in the unresolved region are added in numerically, keeps all three terms [10].

In Chap. 9, studying dijet phenomenology at $\sqrt{s} = 14\,\text{TeV}$, we will show that NLOJET++ (using the subtraction method) and JETRAD (using the phase space slicing method) generate consistent results for the region of phase space that has been investigated (see e.g. Figs 9.1 and 9.2).

References

1. E.W.N. Glover, W.T. Giele, D.A. Kosower, Higher order corrections to jet cross sections in hadron colliders. Nucl. Phys. **B403**, 633 (1993) (hep-ph/9302225)
2. R.K. Ellis, D.A. Ross, A.E. Terrano, The perturbative calculation of jet structure in e^+e^- annihilation. Nucl. Phys. B **178**, 421–456 (1981)
3. S. Catani, M.H. Seymour, A general algorithm for calculating jet cross sections in NLO QCD. Nucl. Phys. **B485**, 291–419 (1997)
4. Z. Nagy, Next-to-leading order calculation of three-jet observables in hadron–hadron collision. Phys. Rev. **D68**, 094002 (2003) (hep-ph/0307268)
5. T. Kinoshita, Mass singularities of feynman amplitudes. J. Math. Phys. **3**, 650–677 (1962)
6. T.D. Lee, M. Nauenberg, Degenerate systems and mass singularities. Phys. Rev. **133**(6B), 1549–1562 (1964)
7. N. Nakanishi, Prog. Theor. Phys. **19**, 159 (1958)
8. L.G. Peter, VEGAS: an adaptive multidimensional integration program. CLNS **80**, 447 (1980)
9. W.B. Kilgore, W.T. Giele, Next-to-leading order gluonic three jet production at hadron colliders. Phys. Rev. D **55**, 7183 (1997) (hep-ph/9610433)
10. Z. Kunszt, D.E. Soper, Calculation of jet cross-sections in hadron collisions at order alpha-s**3. Phys. Rev. **D46**, 192–221 (1992)

Chapter 4
Dijet Physics at Colliders

4.1 Leading-Order Jet Pair Production

4.1.1 Massless Partons

In QCD, two-jet events result when an incoming parton from one hadron scatters off an incoming parton from the other hadron and produces two high transverse momentum partons which are observed as jets. From momentum conservation, the two final-state partons are produced with equal and opposite momenta in the center-of-mass frame of the subprocess. If only two partons are produced, and the relatively small intrinsic transverse momentum of the incoming partons is neglected, the two jets will also be back-to-back in azimuth and balanced in transverse momentum in the laboratory frame.

Below we will discuss cross sections using the Mandelstam variables. For a $2 \rightarrow 2$ process, symbolically $1 + 2 \rightarrow 3 + 4$, we define:

$$\hat{s} \equiv (p_1 + p_2)^2, \tag{4.1}$$

$$\hat{t} \equiv (p_1 - p_3)^2 = -\frac{1}{2}\hat{s}\big(1 - \cos(\hat{\theta})\big), \tag{4.2}$$

$$\hat{u} \equiv (p_2 - p_3)^2 = -\frac{1}{2}\hat{s}\big(1 + \cos(\hat{\theta})\big), \tag{4.3}$$

with $\hat{\theta}$ the center-of-mass scattering angle and $\hat{s}$ the squared center-of-mass energy in the parton–parton frame ($\hat{s} = x_1 x_2 s$, with $x_{1,2}$ defined in the same way as in Eq. 2.19). The equality signs in Eqs. 4.2 and 4.3 are only correct in the case of massless particles. Also, note that in the case of massless particles $\hat{s} + \hat{t} + \hat{u} = 0$.

In this section, we approximate partons as being massless, the next section discusses the modifications for non-negligible masses. The cross section for the following $2 \rightarrow 2$ parton scattering process:

$$\text{parton}_i(p_1) + \text{parton}_j(p_2) \rightarrow \text{parton}_k(p_3) + \text{parton}_l(p_4), \tag{4.4}$$

N. Boelaert, *Dijet Angular Distributions in Proton–Proton Collisions*, Springer Theses,
DOI: 10.1007/978-3-642-24597-8_4, © Springer-Verlag Berlin Heidelberg 2012

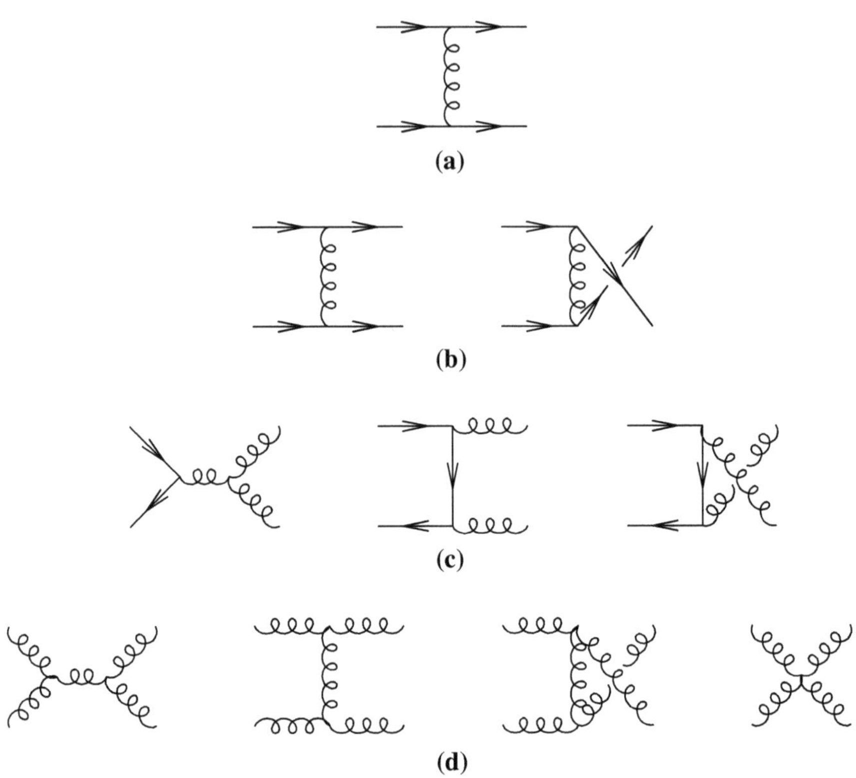

Fig. 4.1 Some of the diagrams for jet production, the other diagrams can be obtained by crossing. The distinction is made between identical (rows **b**, **c** and **d**) and non-identical (row **a**) final-state partons. Figure taken from Ref. [1]

is given by

$$\frac{\mathrm{d}\hat{\sigma}}{\mathrm{d}\hat{t}} = \frac{1}{2\hat{s}^2}\frac{1}{8\pi}\overline{\sum}|\mathcal{M}|^2\delta^4(p_1 + p_2 - p_3 - p_4), \tag{4.5}$$

where $\overline{\sum}$ denotes the average and sum over the initial- and final-state spins and colors respectively. Figure 4.1 shows some of the leading-order diagrams contributing to this cross section [1], the other diagrams can be obtained by crossing. The distinction is made between identical and non-identical final-state partons.

The leading-order expressions for the matrix elements squared $\overline{\sum}|\mathcal{M}|^2$, together with their value at a 90° scattering angle in the parton–parton center-of-mass system ($\hat{\theta} = \pi/2, \hat{t} = \hat{u} = -\hat{s}/2$) are shown in Table 4.1.

Numerically, the most important subprocesses for a pp collider are $qq' \to qq'$, $gq \to qg$ and $gg \to gg$. Let us first concentrate on $qq' \to qq'$. From Table 4.1, the matrix element squared for this process behaves as $\alpha_s^2 \frac{\hat{s}^2+\hat{u}^2}{\hat{t}^2}$. Because $|\hat{t}| < \hat{s}$ and $|\hat{u}| < \hat{s}$, it follows that:

Table 4.1 The invariant matrix elements squared $\overline{\sum}|\mathcal{M}|^2$ for $2 \rightarrow 2$ parton subprocesses with massless partons. The color and spin indices are averaged (summed) over initial (final) states

Process	$\overline{\sum}\|\mathcal{M}\|^2/g^4$	$\hat{\theta}$
$qq' \rightarrow qq'$	$\frac{4}{9}\frac{\hat{s}^2+\hat{u}^2}{\hat{t}^2}$	2.22
$q\bar{q}' \rightarrow q\bar{q}'$	$\frac{4}{9}\frac{\hat{s}^2+\hat{u}^2}{\hat{t}^2}$	2.22
$qq \rightarrow qq$	$\frac{4}{9}\left(\frac{\hat{s}^2+\hat{u}^2}{\hat{t}^2}+\frac{\hat{s}^2+\hat{t}^2}{\hat{u}^2}\right)-\frac{8}{27}\frac{\hat{s}^2}{\hat{u}\hat{t}}$	3.26
$q\bar{q} \rightarrow q'\bar{q}'$	$\frac{4}{9}\frac{\hat{t}^2+\hat{u}^2}{\hat{s}^2}$	0.22
$q\bar{q} \rightarrow q\bar{q}$	$\frac{4}{9}\left(\frac{\hat{s}^2+\hat{u}^2}{\hat{t}^2}+\frac{\hat{t}^2+\hat{u}^2}{\hat{s}^2}\right)-\frac{8}{27}\frac{\hat{u}^2}{\hat{s}\hat{t}}$	2.59
$q\bar{q} \rightarrow gg$	$\frac{32}{27}\frac{\hat{t}^2+\hat{u}^2}{\hat{t}\hat{u}}-\frac{8}{3}\frac{\hat{t}^2+\hat{u}^2}{\hat{s}^2}$	1.04
$gg \rightarrow q\bar{q}$	$\frac{1}{6}\frac{\hat{t}^2+\hat{u}^2}{\hat{t}\hat{u}}-\frac{3}{8}\frac{\hat{t}^2+\hat{u}^2}{\hat{s}^2}$	0.15
$gq \rightarrow gq$	$-\frac{4}{9}\frac{\hat{s}^2+\hat{u}^2}{\hat{s}\hat{u}}+\frac{\hat{u}^2+\hat{s}^2}{\hat{t}^2}$	6.11
$gg \rightarrow gg$	$\frac{9}{2}\left(3-\frac{\hat{t}\hat{u}}{\hat{s}^2}-\frac{\hat{s}\hat{u}}{\hat{t}^2}-\frac{\hat{s}\hat{t}}{\hat{u}^2}\right)$	30.4

$$\frac{\hat{s}^2}{\hat{t}^2} < \frac{\hat{s}^2+\hat{u}^2}{\hat{t}^2} < \frac{2\hat{s}^2}{\hat{t}^2}, \tag{4.6}$$

and thus in a good approximation, $\frac{\hat{s}^2+\hat{u}^2}{\hat{t}^2} \approx \frac{\hat{s}^2}{\hat{t}^2}$. For the other processes, if singular, it can be shown that there is a similar behavior for the matrix elements squared when $\hat{t} \rightarrow 0$, i.e. for small scattering angles when the t-channel gluon exchanges are dominating. Therefore, in a good approximation the parton cross section is approximately inversely proportional to $\hat{t}^2$:

$$\frac{\mathrm{d}\hat{\sigma}}{\mathrm{d}\hat{t}} \propto \frac{\alpha_s^2}{\hat{t}^2} \tag{4.7}$$

We will now discuss the kinematics in more detail. For this purpose, rapidity y is introduced:

$$y = \frac{1}{2}\ln\left(\frac{E+p_z}{E-p_z}\right), \tag{4.8}$$

so that the four-momentum of a massless particle is given by:

$$\begin{aligned} p^\mu &= (E, p_x, p_y, p_z) \\ &= \big(p_T\cosh(y), p_T\sin(\phi), p_T\cos(\phi), p_T\sinh(y)\big) \end{aligned} \tag{4.9}$$

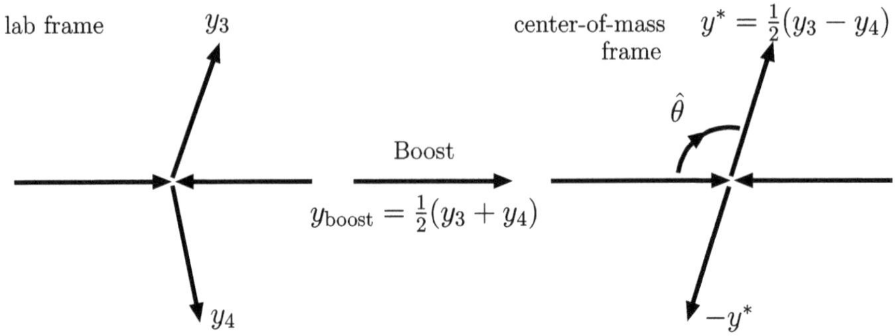

Fig. 4.2 Two massless partons in the lab frame boosted to the center-of-mass frame giving them equal but opposite rapidities

Note that the rapidity difference between two particles is invariant for boosts along the z-axis. We also introduce $y_{\text{boost}} = (y_3 + y_4)/2$ and $y^* = (y_3 - y_4)/2$. A longitudinal boost y_{boost} will boost the lab to the center-of-mass frame, giving the particles equal but opposite rapidities. Figure 4.2 shows what is going on. For massless partons, the center-of-mass scattering angle $\hat{\theta}$ is given by $\cos(\hat{\theta}) = \tanh(y^*)$.

The colliding partons carry a fraction $x_{1,2}$ of the momentum of the hadron they belong to, and, using momentum conservation, the momentum fractions can be determined as:

$$\begin{aligned} x_1 &= \frac{1}{2} x_T \left(e^{y_3} + e^{y_4} \right) = x_T e^{y_{\text{boost}}} \cosh(y^*) \\ x_2 &= \frac{1}{2} x_T \left(e^{-y_3} + e^{-y_4} \right) = x_T e^{-y_{\text{boost}}} \cosh(y^*), \end{aligned} \tag{4.10}$$

where $x_T = 2p_T/\sqrt{s}$.

At the LHC ($\sqrt{s} = 14$ TeV), a central hard jet of 100 GeV corresponds typically to $x \approx 0.01$. Figure 2.2 shows that this is the region where the gluon PDF is big. This means that the LHC is actually a gluon–gluon collider.

The invariant mass M_{jj} of the jet–jet (parton–parton) system is given by:

$$M_{jj}^2 = \hat{s} = 4p_T^2 \cosh^2(y^*) \tag{4.11}$$

4.1.2 Massive Particles

Above formulae are valid at lowest order, and for massless particles. The four-momentum of a massive particle with mass m is no longer given by Eq. 4.9, but is written as

$$p^\mu = \left(m_T \cosh(y), \, p_T \sin(\phi), \, p_T \cos(\phi), \, m_T \sinh(y) \right), \tag{4.12}$$

with the transverse mass $m_T = \sqrt{p_T^2 + m^2}$.

Consequently, the invariant mass M_{jj} between two massive particles, is given by:

$$M_{jj}^2 = m_1^2 + m_2^2 + 2m_{T,1}m_{T,2}\cosh(y_1 - y_2) - 2p_{T,1}p_{T,2}\cos(\phi_1 - \phi_2) \quad (4.13)$$

Experimentally, the mass of a jet is not well defined, and therefore pseudorapidity is preferred over rapidity:

$$\eta = \frac{1}{2}\ln\left(\frac{|p| + p_z}{|p| - p_z}\right) = -\ln\left[\tan\left(\frac{\theta}{2}\right)\right], \quad (4.14)$$

with θ the angle the particle is making with the beam axis. The longitudinal momentum is given by $p_z = p_T\sinh(\eta)$, while the total three-momentum, $|\mathbf{p}| = \sqrt{p_x^2 + p_y^2 + p_z^2}$, is given by $|\mathbf{p}| = p_T\cosh(\eta)$. The pseudorapidity approaches the true rapidity in the massless limit. In hard scattering processes, the masses can usually be safely neglected and $\eta \simeq y$.

The transverse energy $E_T = E\sin(\theta)$ is defined as the projection of the energy onto the transverse direction of the beam. Because $\cosh(\eta) = 1/\sin(\theta)$ (and $\sinh(\eta) = \cos(\theta)/\sin(\theta)$), the energy E can be rewritten as $E_T\cosh(\eta)$, and the four-momentum takes the following form:

$$p^\mu = \big(E_T\cosh(\eta), p_T\sin(\phi), p_T\cos(\phi), p_T\sinh(\eta)\big) \quad (4.15)$$

4.2 Dijet Angular Distributions

4.2.1 Parton Level Considerations

Equation 4.7 rewritten in terms of $\hat{\theta}$ reveals that this cross section corresponds to Rutherford scattering, i.e. a cross section that peaks at small scattering angles:

$$\mathrm{d}\hat{\sigma}/\mathrm{d}(\cos\hat{\theta}) \propto \sin^{-4}(\hat{\theta}/2), \quad (4.16)$$

which is common to all t-channel exchanges of massless vector bosons (gluons and photons). Experimentally, the study of the angular behavior is done using the variable χ, which is defined as

$$\chi = \exp\left(|\eta_1 - \eta_2|\right) = \exp\left(2|\eta^*|\right), \quad (4.17)$$

with η_1 and η_2 the pseudorapidities of the two hardest jets in the event and $\eta^* = \frac{1}{2}(\eta_1 - \eta_2)$. For massless particles, i.e. when the pseudorapidity equals the true rapidity, χ is invariant under Lorentz boosts along the beam axis and can be calculated in the rest frame of the experiment.

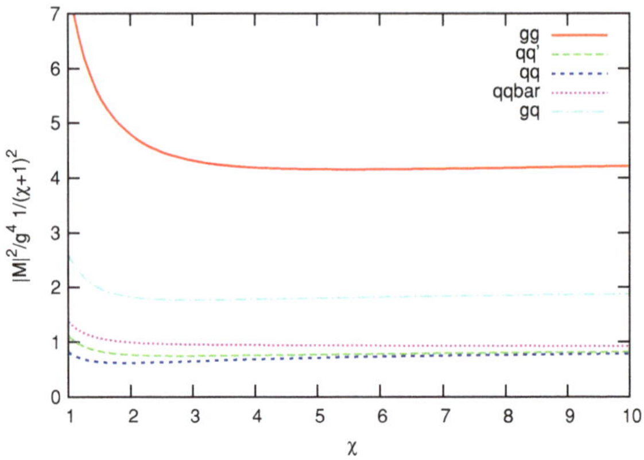

Fig. 4.3 Subprocess matrix elements χ-dependence, $\frac{|\mathcal{M}|^2}{g^4}\frac{1}{(\chi+1)^2}$, for $\hat{s} = 1\,\text{GeV}^2$

At lowest order, i.e. for a $2 \rightarrow 2$ process, the pseudorapidity of the two massless particles in the center-of-mass frame is given by $\pm\eta^*$ and can be related to the scattering angle $\hat{\theta} = \arccos(\tanh \eta^*)$, and therefore also the following expression for χ as a function of $\hat{\theta}$ holds:

$$\chi = \frac{1 + |\cos\hat{\theta}|}{1 - |\cos\hat{\theta}|} \sim \frac{1}{1 - |\cos\hat{\theta}|} \propto \frac{\hat{s}}{\hat{t}} \tag{4.18}$$

Using the approximation that $\chi \propto \hat{s}/\hat{t}$ and keeping $\hat{s}$ fixed, the partonic cross section in Eq. 4.7 can now be rewritten as a function of χ and turns out to be approximately constant:

$$\frac{\mathrm{d}\hat{\sigma}}{\mathrm{d}\chi} \propto \frac{\alpha_s^2}{\hat{s}} \quad (\hat{s}\,\text{fixed}) \tag{4.19}$$

On the other hand, a cross section that is isotropic in $\hat{\theta}$, meaning that $\mathrm{d}\hat{\sigma}/\mathrm{d}(\cos\hat{\theta}) \propto$ constant, will give rise to a cross section in χ that is peaked at low χ values, i.e. $\frac{\mathrm{d}\hat{\sigma}}{\mathrm{d}\chi} \propto 1/(\chi + 1)^2$. New physics is typically hidden in isotropic events; this will be shown later on (see e.g. Sects. 5.3.4 and 5.4.1).

Using $\cos\hat{\theta} = (\chi - 1)/(\chi + 1)$, the explicit χ-dependence of several inclusive subprocesses is plotted in Fig. 4.3 for $\hat{s} = 1\,\text{GeV}^2$. The individual subprocesses have been symmetrized in $\hat{t}$ and $\hat{u}$ for non-identical particles in the final state. Because $\mathrm{d}t/\mathrm{d}\chi = 2\hat{s}/(\chi + 1)^2$, the matrix elements are scaled with $(\chi + 1)^{-2}$ in order to display the correct behavior. Note that all contributions are quite similar in shape—i.e. flat for a large region in χ—and that the *gg* subprocess is by far the most dominant. The fact that the cross section for an isotropic event drastically differs in shape from QCD scattering in the case of fixed $\hat{s}$, makes that this cross section measurement is preferred to be done in bins of $\hat{s}$. In terms of χ, $\hat{s}$ (Eq. 4.11) can be rewritten as:

$$\sqrt{\hat{s}} = M_{jj} = p_T \left(\sqrt{\chi} + \frac{1}{\sqrt{\chi}}\right) = 2p_T \cosh\left(\log(\chi)/2\right) \tag{4.20}$$

The above equations tell us that, for $\hat{s}$ fixed, low values of χ correspond to high values of p_T and vice versa.

4.2.2 Hadron Level Considerations

To obtain the result at the hadron level, the partonic cross section needs to be multiplied with the parton distribution functions and integrated over the momentum fractions (see also Eq. 2.19):

$$\frac{\mathrm{d}\sigma}{\mathrm{d}\chi} = \int \mathrm{d}x_1 \int \mathrm{d}x_2 f_1(x_1, Q^2) f_2(x_2, Q^2) \frac{\mathrm{d}\hat{\sigma}}{\mathrm{d}\chi} \tag{4.21}$$

This differential cross section $\mathrm{d}\sigma/\mathrm{d}\chi$ vs χ is commonly referred to as the dijet angular distribution. Because at lowest order the dijet invariant mass M_{jj} equals $\sqrt{\hat{s}}$, this measurement is done in bins of M_{jj}. Apart from binning in dijet invariant mass, also selection cuts in pseudorapidity are made in order to improve the QCD sensitivity of this measurement. We will discuss them.

At lowest order, the variables determining the cross section in Eq. 4.21 are x_1, x_2 and χ. Note also that $\mathrm{d}x_1\mathrm{d}x_2$ can be rewritten as $\mathrm{d}\tau\mathrm{d}y$, where $\tau = x_1x_2 = \hat{s}/s$ and $y = (1/2)\ln(x_1/x_2) = (\eta_1 + \eta_2)/2$. This means that an equivalent set of parameters is $\hat{s}$, $(\eta_1 + \eta_2)/2$ and χ, with the partonic cross section depending on $\hat{s}$ and χ only.

In Sect. 2.4 it was mentioned that the parton distribution function only shows logarithmic QCD scaling violations (up to leading order) and from itself does not carry a direct QCD prediction. On the other hand, the partonic cross section is a direct prediction of QCD. This means that, if we want to use Eq. 4.21 to study QCD, we should try to keep x_1 and x_2 fixed and study the remaining variation in χ. But this ideal view is not a realistic one because of statistical reasons, and therefore we need to open up the range in x_1 and x_2.

Keeping $\hat{s}$ fixed implies that in a given bin of χ only $(\eta_1 + \eta_2)/2$ can vary. In this bin, for each $(\eta_1 + \eta_2)/2$ value, the same partonic cross section is probed [because it has no dependence on $(\eta_1 + \eta_2)/2$], but the sampling points are weighted with different PDF factors. In order to probe the partonic cross section over the whole range of χ and not the PDF convolution, the same $(\eta_1 + \eta_2)/2$ range in each bin of χ has to be used, otherwise different χ bins cannot be easily compared. In other words, a selection cut of the format $|\eta_1 + \eta_2|/2 < (1/2)c$ is required.

Experimentally, a detector measures jets up to a certain η_{max}. In order to benefit maximally from the detector, the following orthogonal selection cuts need to be made for an angular distribution measurement up to χ_{max}:

$$|\eta_1 + \eta_2| < c, |\eta_1 - \eta_2| < 2\eta_{\mathrm{max}} - c, \tag{4.22}$$

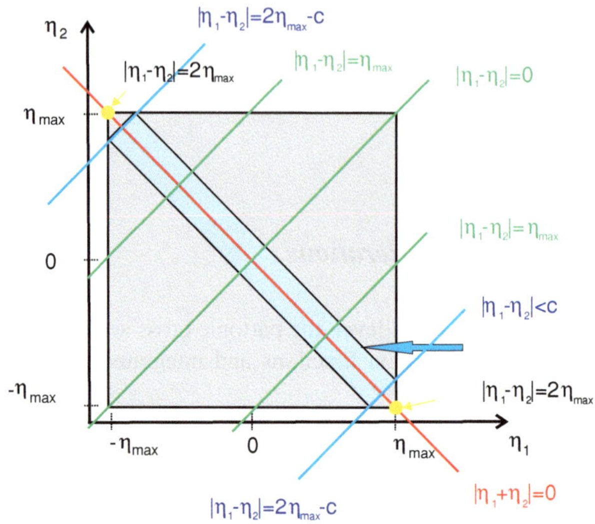

Fig. 4.4 Pseudorapidity range of a generic detector, before (*grey square*) and after (*blue rectangle*) applying selection cuts Eq. 4.22. The lines parallel to the diagonal are lines of constant $|\eta_1 - \eta_2|$

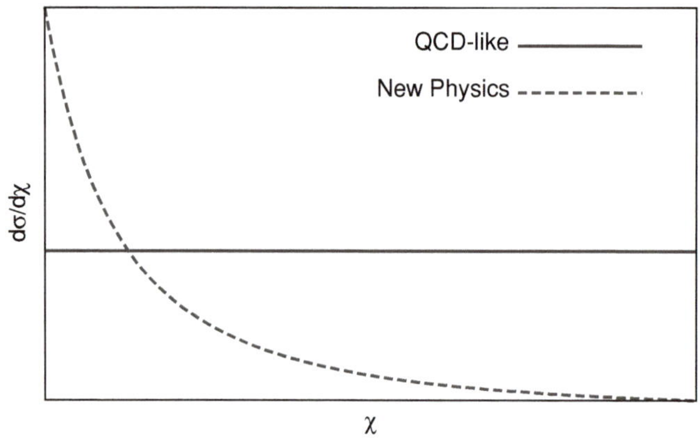

Fig. 4.5 Dijet angular distributions from QCD and new physics are expected to behave differently at low values of χ; the QCD curve tends to be flat, while new physics curves are more peaked at low χ because they usually contain more isotropic processes. The exact behavior of the QCD parton cross section is shown in Fig. 4.3

where $\chi_{\text{max}} = \exp(2\eta_{\text{max}} - c)$, and the parameter c is a trade-off between statistics, sensitivity to the partonic cross section and χ-range. Figure 4.4 illustrates this for a generic detector; the axes represent the pseudorapidity reaches of jet 1 and jet 2, the grey square is the combined reach in pseudorapidity of the detector, and the blue rectangle shows what is left after applying the selection cuts in Eq. 4.4. The values of η_{max} and c will be discussed in later chapters.

4.3 Conclusion

To summarize, in order to mainly probe the partonic cross section and to a lesser extent the PDFs, the dijet angular distribution $d\sigma/d\chi$ vs χ is measured in bins of dijet invariant mass and with cuts in pseudorapidity as defined in Eq. 4.22. With these cuts, lowest order QCD tells us that the distribution is flat to a large extent. On the other hand, the dijet angular distribution for processes that are more isotropic will be peaked at small χ, or equivalently high p_T. In later chapters we will show that this is often the case for new physics. This different behavior makes the dijet angular distribution measurement a suitable tool to probe new physics, see Fig. 4.5. Measurements carried out in the past were published in references [2–11].

References

1. R.W. Ellis, W.J. Stirling, B.R. Webber, *QCD and Collider Physics*. Cambridge Monographs on Particle Physics, Nuclear Physics and Cosmology. (Cambridge University Press, Cambridge, 1996)
2. UA1 Collaboration CERN, G. Arnison et al., Angular distributions and structure functions from two-jet events at the CERN SPS collider. Phys. Lett. B **136**(4), 294–300 (1984)
3. UA1 Collaboration CERN, G. Arnison et al., Comparison of three-jet and two-jet cross sections in pp collisions at the CERN SPS pp collider. Phys. Lett. B **158**(6), 494–504 (1985)
4. UA1 Collaboration CERN, G. Arnison et al., Angular distributions for high-mass jet pairs and a limit on the energy scale of compositeness for quarks from the CERN pp collider. Phys. Lett. B **177**(2), 244–250 (1986)
5. UA2 Collaboration CERN, P. Bagnaia et al., Measurement of jet production properties at the CERN Collider. Phys. Lett. B **144**, 283–290 (1984)
6. CDF Collaboration, F. Abe et al., Dijet angular distributions from $p\bar{p}$ collisions at $\sqrt{s} = 1.8$ TeV. Phys. Rev. Lett. **62**(26), 3020–3023 (1989)
7. CDF Collaboration, F. Abe et al., Dijet angular distribution in $p\bar{p}$ collisions at $\sqrt{s} = 1.8$ TeV. Phys. Rev. Lett. **69**(20), 2896–2900 (1992)
8. CDF Collaboration, F. Abe et al., Measurement of Dijet Angular Distributions by the Collider Detector at Fermilab. Phys. Rev. Lett. **77**(27), 5336–5341 (1996)
9. CDF Collaboration, Dijet angular distribution in $p\bar{p}$ collisions at $\sqrt{s} = 1.8$ TeV CDF/ANAL/JET/PUB/, (2008), p. 9609
10. D0 Collaboration, Measurement of dijet angular distributions at $\sqrt{s} = 1.96$ TeV and searches for quark compositeness and extra spatial dimensions. Phys. Rev. Lett. **103**, 191803 (2009) (hep-ex/09064819)
11. D0 collaboration, B. Abbott et al., Measurement of Dijet Angular Distributions and Search for Quark Compositeness. Phys. Rev. Lett. **80**(4), 666–671 (1998)

Chapter 5
Gravitational Scattering and Black Holes in Large Extra Dimensions

5.1 Extra Dimensions

Even though the Standard Model is fairly complete, some phenomena cannot be explained with this theory, such as the fact that gravity appears to be much weaker than all other forces. In the 1920s, Kaluza and Klein posted the idea that the existence of extra spatial dimensions provides a mechanism to unify gravity with electromagnetism. Even though this theory failed in its original purpose due to internal inconsistencies, the idea of extra dimensions was set and became rather popular in the second half of the twentieth century. Nowadays it forms the basis of string theory.

The fact that extra dimensions have not been observed yet can be explained by assuming that they are compactified to a small scale, with a compactification radius R. E.g. for one extra dimension, we can imagine in each spacetime point a circle with radius R (cfr. Fig. 5.1). Spacetime then has the topology $\mathcal{M}_4 \times S_1$, with $\mathcal{M}_4$ the ordinary four-dimensional Minkowski spacetime. When R is sufficiently small, we cannot probe this circle in our every day world, and we cannot directly feel the effect of the extra dimension.

At first, the size of the extra dimensions was dictated by the inverse of the observed four-dimensional Planck Scale ($\sim 10^{19}$ GeV), but more modern models such as the ADD [1, 2], see Sect. 5.2, allow for values of R as large as submillimeter.

5.1.1 Kaluza-Klein Mode Expansion and Reduction

In the general case of n extra dimensions, spacetime has the factorized topology $\mathcal{M}_4 \times \mathcal{K}$, with $\mathcal{K}$ an n-dimensional compact manifold.[1] The fact that $\mathcal{K}$ is compact means that the coordinates on the volume obey boundary conditions. Going back to

[1] Even though non-factorizable geometries exist and give rise to interesting physics, we will not discuss them here.

N. Boelaert, *Dijet Angular Distributions in Proton–Proton Collisions*, Springer Theses, DOI: 10.1007/978-3-642-24597-8_5, © Springer-Verlag Berlin Heidelberg 2012

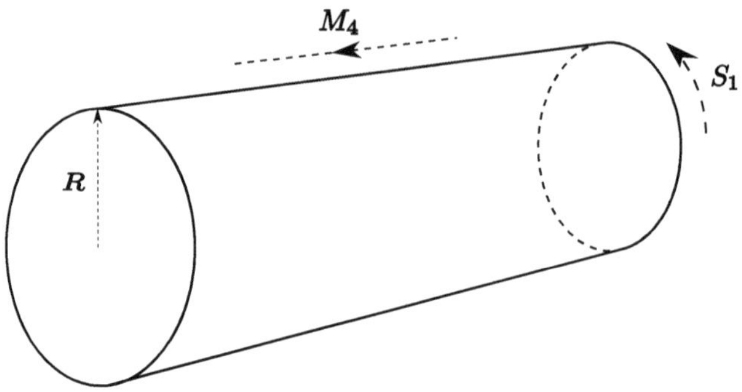

Fig. 5.1 One extra dimension compactified on a circle of radius R. At every point in the four-dimensional spacetime (here collapsed onto a line), there exists an "orthogonal" dimension compactified on a circle of radius R. This way, spacetime has the topology $\mathcal{M}_4 \times S_1$

the example of the one-dimensional circle, the coordinate y describing the circle is periodic with periodicity $2\pi R$.

In a $D = 4 + n$-dimensional world, the D-dimensional quantum fields, which are representations of the D-dimensional Lorentz group, are used to formulate the D-dimensional action S_D. The fields can be decomposed into a sum of irreducible representations of the four-dimensional Lorentz group, which will yield a spectrum of fields of different four-dimensional spins. Next, a mode expansion for each field can be performed, taking into account the boundary conditions. In the case of the one-dimensional circle, the most general expansion of the wavefunction $\Phi(x^\mu, y)$, with x^μ the coordinates on $\mathcal{M}_4$ and y the coordinate on S_1, consistent with the compactification symmetry, takes the following form [3]:

$$\Phi(x^\mu, y) = \frac{1}{\sqrt{2\pi R}} \sum_{n=-\infty}^{\infty} \phi_n(x^\mu) \exp(iny/R) \tag{5.1}$$

In the general case, the exponentials in Eq. 5.1 are replaced by mutually orthogonal eigenfunctions of the Laplace operator on the manifold $\mathcal{K}$.

We then insert the mode expansion into the D-dimensional action S_D, and integrate S_D over the volume $\mathcal{K}$ in order to obtain the four-dimensional action S_4:

$$S_4 = \int_{\mathcal{K}} \mathrm{d}^n y S_D \tag{5.2}$$

In the example above, imagine that $\Phi(x^\mu, y)$ is a complex Klein-Gordon field of mass m_0, with the five-dimensional action given by

$$S_5 = \int \mathrm{d}^4 x \int_0^{2\pi R} \mathrm{d}y \left[\frac{1}{2} (\partial_M \Phi)^* (\partial^M \Phi) - \frac{1}{2} m_0^2 \Phi \Phi^* \right], \tag{5.3}$$

where M is the five-dimensional spacetime index. After inserting the mode expansion and taking into account the orthogonality of two different modes, the four-dimensional action becomes:

$$S_4 = \int \mathrm{d}^4x\mathrm{d}y \Big\{ \frac{1}{2} \sum_n (\partial_\mu \phi_n)^* (\partial^\mu \phi_n) - \frac{1}{2} \sum_n [m_0^2 + \frac{n^2}{R^2}] \phi_n \phi_n^* \Big\} \tag{5.4}$$

What we get is an infinite tower of Klein-Gordon fields, ϕ_n, with masses given by:

$$m^2 = m_0^2 + \frac{n^2}{R^2} \tag{5.5}$$

We can interpret $\frac{n}{R}$ as the—quantized—momentum in the fifth dimension, which is then absorbed by the effective mass in the effective four-dimensional theory. This is called *Kaluza-Klein (KK) reduction*, which is explained here for a simple case, but can be generalized to other fields and actions. The lightest field is obtained for $n = 0$ and is called the KK zero mode or ground state, while the other fields are called excited KK-states. Excited states have exactly the same quantum numbers (spins, couplings) as those of the ground state mode.

It generally holds that four-dimensional KK-mass terms emerge from kinetic terms in the higher-dimensional Lagrangian. From Eq. 5.5, we note that the threshold energy for detecting excited states is of the order R^{-1}, since this is the threshold for producing the first excited state.

5.1.2 Compactification on an Orbifold

In order to describe better the reality, compactification is done rather on an orbifold than on a manifold. An orbifold is a manifold with some special points, caused by a superimposed discrete symmetry Γ. The discrete symmetry eliminates certain points of the manifold $\mathcal{K}$, and the resulting space $\mathcal{K}/\Gamma$ is called an orbifold. E.g. the discrete $\mathbb{Z}_2$ identification $\Gamma: y \leftrightarrow -y$ in our example would change our circle to a line segment of length πR. The original domain $0 \le y < 2\pi R$ is reduced to $0 \le y < \pi R$, and, unlike the case of the circle, the endpoints of this smaller domain are fixed and not identified with each other. The resulting geometry is $S_1/\mathbb{Z}_2$.

Orbifold compactification does not affect the general idea of KK reduction, but since the higher-dimensional actions must be invariant under this discrete symmetry, certain interaction terms coming from higher dimensions will be removed in the KK-reduction, which allows the theory to describe much better low energy phenomenology. Imposing the symmetry on the fields will reduce the number of excited levels in the KK-expansion of the fields (Eq. 5.1) as well, since the original KK mode expansion has to be rewritten in terms that are also eigenfunctions under Γ, followed by an elimination step in which only those modes that belong to a specific eigenvalue of Γ are kept, removing certain degrees of freedom of the theory. Fields that transform non-trivially under the orbifold discrete action will have their zero-mode eliminated.

5.1.3 Types of Extra Dimensions

There are two big classes of extra dimensions. In *universal* extra dimensions, all forces and particles feel the extra dimensions.

However, there also exist Dirichlet branes (D-branes), dynamical, fluctuating hypersurfaces of dimension p that float in the D-dimensional world, capable of trapping certain gauge fields on their surface. Traditionally p represents the number of space dimensions ($p \leq D-1$), rather than spacetime, and one speaks of a Dp-brane. Extra dimensions in the bulk are called *transverse*, while extra dimensions on the brane (for $p > 3$) are called *longitudinal.*

A common case is the D3-brane that is capable of trapping *all* gauge forces. All gauge bosons and all particles that carry gauge charges are restricted to the brane, and do not accrue KK-excitations. But states that do not feel these forces, such as the graviton, are free to wander off the brane and experience the full D-dimensional world. For this reason, this type of extra dimensions is often called *gravity-only* extra dimensions, and its existence forms the basis of the ADD-model which will be the topic of Sect.5.2.

5.1.4 Bounds on Extra Dimensions

At the present time, there exists no experimental evidence for large extra dimensions; so far, the (lightest) Kaluza-Klein excitations of particles have not been discovered, nor have been indirect effects such as loops or additional effective interactions.

Instead, since collider experiments have probed the structure of matter down to a length scale near the TeV range, without detecting any extra dimension, we can set a limit on the size of *uniform* extra dimensions:

$$R^{-1} \geq \mathcal{O}(1)\,\mathrm{TeV} \tag{5.6}$$

Experiments conducted to test the gravitational inverse-square law can be used to set bounds on the existence of *gravity-only* extra dimensions. In the normal four-dimensional world, the graviton is responsible for ordinary Newtonian gravity, with the inverse-square law arising from graviton exchange, and a potential given by $V(r) = -G_N^{(4)} M_1 M_2 / r$.

However, if we do not have a single graviton, but a KK-tower of gravitons, with the ordinary graviton the KK zero-mode, we must consider the exchanges of the excited states as well, and sum over them. This way the potential changes to:

$$V(r) = -\frac{G_N^{(4)} M_1 M_2}{r}\Big[1 + \sum_{m_n} g_{m_n} e^{-m_n r}\Big], \tag{5.7}$$

with g_{m_n} the graviton state for the nth excited level and $m_n = n/R$. For distances $r \gg R$ all modes except $n = 1$ are suppressed.

Recent experiments have found that the inverse-square law holds down to a length $\lambda = 56\ \mu\text{m}$ [4]. A violation of the inverse-square law is parametrized as

$$V(r) = -\frac{G_N^{(4)} M_1 M_2}{r}\Big[1 + \alpha e^{-r/\lambda}\Big], \tag{5.8}$$

with $\alpha = 8n/3$ and λ the Compton Wavelength of the first KK mode, equal to the radius R. For $n = 2$ (1), one has that

$$R \le 37(44)\ \mu\text{m} \Longleftrightarrow R^{-1} \ge \mathcal{O}(10^{-13})\ \text{TeV} \tag{5.9}$$

Gravity can also be tested indirectly through cosmology or astrophysics, but these measurements do not manage to further constrain the limit in Eq. 5.9.

5.2 The ADD-Model

5.2.1 Concept

An attempt to address the hierarchy problem in particle physics, i.e. why the electroweak scale is so many orders of magnitude lower than the observed Planck scale, was done by Arkani-Hamed, Dimopoulos and Dvali in the late nineties of the twentieth century [1, 2]. Their so-called ADD model assumes the existence of *gravity-only* dimensions in which gravity is allowed to propagate, while the SM fields are confined to a four-dimensional membrane (see Sect. 5.1.3). Figure 5.2 shows this schematically.

This way the D-dimensional fundamental Planck scale can be much lower than the observed four-dimensional one, allowing the fundamental energy scale associated with quantum gravity to be in the TeV range, of the same order of the electroweak scale.

5.2.2 Lowering the Planck Scale

In $D = 4 + n$ dimensions, two test masses of mass m_1 and m_2 placed within a distance $r \ll R$ will feel a gravitational potential dictated by Gauss' law:

$$V(r) \sim \frac{m_1 m_2}{M_P^{n+2}} \frac{1}{r^{n+1}},\ (r \ll R), \tag{5.10}$$

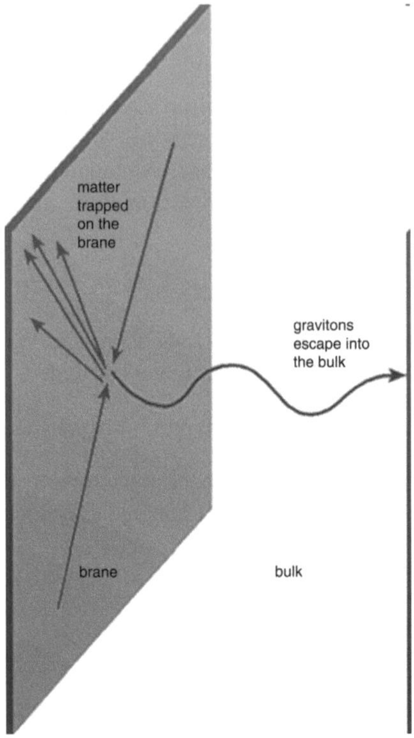

Fig. 5.2 The ADD-model [1, 2] assumes the existence of large extra dimensions in which gravity is allowed to propagate, while SM fields are confined to a four-dimensional membrane.

with M_P the Planck scale in $(n+4)$ dimensions, derived from the $(n+4)$-dimensional Newton's constant:

$$M_P \equiv [G_N^{(n+4)}]^{-1/(n+2)} \tag{5.11}$$

On the other hand, if the masses are placed at distances $r \gg R$, their gravitational flux lines cannot continue to penetrate in the extra dimensions, and the ordinary $1/r$ potential is obtained. Since the potential is a continuous function in r, we have:

$$V(r) \sim \frac{m_1 m_2}{M_P^{n+2} R^n} \frac{1}{r^1}, \; (r \gg R), \tag{5.12}$$

We use the above formula to define the effective four-dimensional Planck scale M_{P4}:

$$M_{P4}^2 \equiv \frac{1}{G_N^{(4)}} \sim M_P^{n+2} R^n, \tag{5.13}$$

so that we obtain the ordinary gravity potential for distances much larger than R:

$$V(r) \sim \frac{m_1 m_2}{M_{P4}^2} \frac{1}{r^1}, \; (r \gg R), \tag{5.14}$$

with $M_{P4} \sim 10^{19}$ GeV.. If reality truly is D-dimensional, then M_{P4} is not a fundamental scale of nature, but M_P is. We note from Eq. 5.13 that since R can take large values, the higher-dimensional Planck scale can remain much smaller than the observed four-dimensional one.

Note that Eqs. 5.7 and 5.10 are consistent with each other, since for $r \ll R$ the sum in Eq. 5.7 can be turned into an integral over $m = |m_{\vec{n}}|$:

$$1 + \sum_{m_n} g_{m_n} e^{-m_n r} \sim \int (R^n m^{n-1}) e^{-mr} \mathrm{d}m \sim \left(\frac{R}{r}\right)^n \sim \frac{M_{P4}^2}{M_P^{n+2}} \left(\frac{1}{r}\right)^n \tag{5.15}$$

The term $(R^n m^{n-1})$ comes from the density of KK states; we will derive this (Eq. 5.22) in Sect. 5.3.1.

In the ADD model, M_P is of the order of 1 TeV. This means that, using Eq. 5.13, one extra dimension ($n = 1$) corresponds to $R \sim 10^{14}$ cm, which is excluded by—daily—experimental observations. The case $n = 2$ corresponds to $R \sim 1$ mm, which is excluded as well (Eq. 5.9). So far, $n > 2$ has not been excluded for fundamental Planck Masses in the TeV range [2]. With the LHC starting up, soon the TeV range will be explored.

5.2.3 Implications for Low Energy Phenomenology

Even though the bounds on extra dimensions are quite generous (Eqs. 5.6 and 5.9), extra dimensions were originally considered to be small. The reason for this was naturalness; their natural scale lies in the neighborhoud of the Planck Scale, giving rise to KK excitations with masses $\sim 10^{19}$ GeV. These excitations will be essentially unobservable, having no notable relevance to the low-energy world.

However, in the previous section, we have seen that if the fundamental Planck Scale is at about 1 TeV, the size of the extra dimensions is no longer extremely small, meaning that the excited KK states are not so remotely heavy. When the KK states become sufficiently light, they must affect the observable physics.

This also means that, for a fundamental Planck scale of around 1 TeV, the ADD model predicts the production of gravity mediated processes at the LHC, with probably as the most striking one the black hole production. But also processes involving the production of on-shell gravitons, or the exchange of virtual Kaluza-Klein (KK) modes of the graviton, with gravitational scattering of hard partons as dominant process, will be present.

In this thesis, we focus on two processes, namely black hole production and gravitational scattering.

5.2.4 Kinematic Regimes

Three different kinematic regimes exist, depending on the scattering energy $\sqrt{\hat{s}}$.

In the *transplanckian region*, where $\sqrt{\hat{s}} \gg M_P$, elastic collisions can be described by linearized general relativity as long as the momentum transfer of the process is sufficiently small. The Schwarzschild radius r_s of a $(4+n)$-dimensional object is given by [5]:

$$r_s = \frac{1}{\sqrt{\pi} M_P} \left[\frac{M_{BH}}{M_P} \frac{8\Gamma(\frac{n+3}{2})}{n+2} \right]^{\frac{1}{n+1}}, \tag{5.16}$$

and the transplanckian regime corresponds to a classical limit in which the length scale r_s characterizes the dynamics. The gravitational field is weak and we will show later on (Sect. 5.3) that the interactions between two colliding partons and the gravitational field can be computed using the eikonal approximation, which can be trusted at small scattering angles. Collisions with an impact parameter (defined as the perpendicular distance between the two colliding objects) smaller than r_s will result in black hole production, see Sect. 5.4.

A classical description of the process is valid if the quantum-mechanical uncertainties in the impact parameter and in the scattering angle are small with respect to their classical values. This is no longer true in the *cisplanckian region*, i.e. the region where $\sqrt{\hat{s}} \ll M_P$. For these low energies, the theory can be described by an effective field theory of graviton KK excitations (Sect. 5.3.1).

In the neighborhood of the fundamental Planck scale, i.e. $\sqrt{\hat{s}} \approx M_P$, quantum gravity effects are no longer negligible. This is the *planckian region*. Unfortunately, since quantum gravity dynamics is still unknown, systematic theoretical predictions cannot be made.

Consequently, if $M_P \approx 1$ TeV, experiments at the LHC could probe all three regions.

5.2.5 Definitions

Unfortunately, the definition of the fundamental Planck scale is not unique but differs from author to author. For clarity we will list the most common ones and relate them to each other.

The 4-dimensional Newtonian constant $G_{N(4)}$ is used to define the 4-dimensional Planck mass M_{P4} [6]:

$$M_{P4} = \frac{1}{\sqrt{G_{N(4)}}} = 1.22 \times 10^{19}\ \text{GeV/c}^2 \tag{5.17}$$

In this thesis we will use the fundamental (higher-dimensional) Planck mass M_P according to Ref. [5, 7, 8]:

$$M_P^{2+n} = \frac{1}{L^n G_{N(4)}} = \frac{M_{P4}^2}{L^n}, \tag{5.18}$$

with the compactification circumference L, related to the compactification radius R: $L = 2\pi R$.

A different definition (and notation) for the fundamental Planck mass M_D was used in Ref. [6]:

$$M_D^{2+n} = \frac{1}{8\pi R^n G_{N(4)}}, \tag{5.19}$$

and thus

$$M_P^{2+n} = 2^{(3-n)}\pi^{(1-n)} M_D^{2+n} \tag{5.20}$$

5.3 Gravitational Scattering in the ADD Model

5.3.1 KK Reduction of the Graviton

A KK reduction of the graviton is presented in Ref. [9, 10]. The theory is formulated as an effective field theory (strictly only valid for $\sqrt{\hat{s}} \ll M_P$) and the starting point for the analysis is the linearized gravity Lagrangian in $D = 4 + n$ dimensions. For simplicity, the fields are compactified on an n-dimensional torus with common radius R. For a given excited KK level $\vec{n}$, there are one spin-2 state, $(n-1)$ spin-1 states and $n(n-1)/2$ spin-0 states, all mass degenerate with mass $m_{\vec{n}}^2 = \frac{\vec{n}^2}{R^2}$. The zero modes, $\vec{n} = 0$, of the $(4+n)$-dimensional graviton become the graviton, n massless $U(1)$ gauge bosons and $n(n+1)/2$ massless scalar bosons.

Since R^{-1} is smaller than the typical energy resolution in collider experiments, the mass distribution of KK states is quasi continuous, and the number of states in the mass interval $\mathrm{d}m_{\vec{n}}^2$ is obtained by

$$\Delta \vec{n}^2 \approx \rho(m_{\vec{n}})\mathrm{d}m_{\vec{n}}^2, \tag{5.21}$$

where the KK state density function, $\rho(m_{\vec{n}})$, is given by:

$$\rho(m_{\vec{n}}) = \frac{\pi^{n/2} R^n m_{\vec{n}}^{n-2}}{\Gamma(n/2)} = \frac{1}{2^n \pi^{n/2}\Gamma(n/2)} \frac{M_{P4}^2}{M_P^{2+n}} m_{\vec{n}}^{n-2} \tag{5.22}$$

This means that the multiplicity of gravitons beneath a relevant energy scale E is $\sim (RE)^n$.

When constructing the effective interactions between KK states and ordinary matter fields, it was found that spin-1 states decouple and spin-0 states only couple through the dilation mode. The interacting Lagrangian for the KK states and SM fields, together with the Feynman rules was presented in Ref. [9, 10].

Each graviton couples to matter with normal (weak) gravitational strength $\sim 1/M_{P4}$, having a negligible effect on particle physics. However, since the density of gravitons beneath a relevant energy scale can be large, the combined effect of all the gravitons is not always negligible. It follows from Eq. 5.22 that gravitational scattering processes are suppressed by powers of $1/M_P$ only.

5.3.2 Scattering Amplitude

In a gravitational scattering event, quantized KK modes occur as intermediate states and the sum over all modes has to be made, which can be turned (see Eq. 5.22) into a divergent propagator integral:

$$\sum_{m_{\vec{n}}} \frac{1}{-m_{\vec{n}}^2 + k^2} \approx \frac{2\pi^{n/2}}{\Gamma(n/2)} R^n \int \frac{m^{n-1}}{-m^2 + k^2} \mathrm{d}m, \tag{5.23}$$

with $\vec{n}$ enumerating the allowed KK modes with momenta $m_{\vec{n}}$, $m = |m_{\vec{n}}|$ and k the exchanged 4-momentum in normal space. This integral diverges for $n \geq 2$, but can be rendered finite by introducing a physical cut-off of the KK tower to be summed over. A cut-off implied from a narrow (small compared to the compactification radius R) width of the four-dimensional membrane was proposed in Ref. [11]. This was done by assuming a Gaussian extension of the standard model field densities into the bulk:

$$\psi(y) = \left(\frac{M_s}{\sqrt{2\pi}}\right)^{\frac{n}{2}} \exp\left(-\frac{y^2 M_s^2}{4}\right), \tag{5.24}$$

with y being the coordinate in the extra dimension. Formula (5.24) introduces a new parameter, M_s, with the dimension of mass, which is a measure for the inverse of the brane width. The overlap between two SM fields and a KK mode of mass m is then proportional to:

$$\int \mathrm{d}y e^{im\cdot y} \left(\frac{M_s}{\sqrt{2\pi}}\right)^{2\frac{n}{2}} \exp\left(-\frac{y^2 M_s^2}{2}\right) = e^{-m^2/(2M_s^2)} \tag{5.25}$$

This suppression factor occurs once at every vertex, so the total suppression for KK exchange will be $e^{-m^2/(M_s^2)}$. In this context M_s can also be regarded as KK mass cut-off.

Instead of Eq. 5.23, we now have an "effective" propagator,

$$\frac{2\pi^{n/2}}{\Gamma(n/2)} R^n \int \frac{m^{n-1}}{-m^2 + k^2} e^{-m^2/M_s^2} \mathrm{d}m, \tag{5.26}$$

which is then used for the Born amplitude:

$$A_{\text{Born}}(k^2 = t) = \frac{\hat{s}^2}{2^{n-3}\pi^{n-1} M_P^{n+2}} S_n \int_0^\infty \frac{m^{n-1}}{-m^2 + k^2} e^{-m^2/M_s^2} \mathrm{d}m, \tag{5.27}$$

in the approximation of ultra-relativistic small angle scattering. This integral is convergent and finite for all negative values of $k^2 = t$, and can be calculated in the limits of large and small momentum transfers. The scattering process now depends on three energy scales, namely the collision energy $\sqrt{\hat{s}}$, the fundamental Planck scale M_P and the inverse of the membrane width M_s.

5.3.3 Large Momentum Transfers: $\sqrt{\hat{s}} \gg M_s$

In the case of large momentum transfers, i.e.$-\hat{t} \gg M_s^2$, the term m^2 in the denominator in Eq. 5.27 can be neglected, which gives the result:

$$A_{\text{Born}}(t) \approx \frac{\hat{s}^2}{2^{n-3}\pi^{n-1} M_P^{n+2}} S_n \int_0^\infty \frac{m^{n-1}}{\hat{t}} e^{-m^2/M_s^2} \mathrm{d}m = \frac{\pi^{n/2}}{2^{n-3}\pi^{n-1}} \left(\frac{M_S}{M_P}\right)^n \frac{\hat{s}^2}{M_P^2 \cdot \hat{t}}, \tag{5.28}$$

which is dominated by small values of $(\hat{t}/\hat{s})$, i.e. small angle scattering via the t-channel. This means that the cross section can be calculated in the eikonal approximation:

$$\frac{\mathrm{d}\sigma_{\text{eik}}}{\mathrm{d}\hat{t}} = \frac{1}{16\pi \hat{s}^2} |A_{\text{eik}}|^2 \tag{5.29}$$

Contributions from multi-loop ladder diagrams exponentiate and the all order eikonal amplitude is given by

$$A_{\text{eik}}(k^2) = -2i\hat{s} \int \mathrm{d}^2 b_T e^{ik_T \cdot b_T} (e^{i\xi} - 1), \tag{5.30}$$

with b the impact parameter and $\xi(b)$ the eikonal scattering phase, given by

$$\begin{aligned} \xi(b) &= \frac{1}{2\hat{s}} \int \frac{\mathrm{d}^2 k_T}{(2\pi)^2} e^{ik_T \cdot b_T} A_{\text{Born}}\left(-k_T^2\right) \\ &= -\frac{\hat{s} M_s^n}{(2\sqrt{\pi})^n M_P^{n+2}} \Gamma\left(\frac{n}{2}\right) U\left(\frac{n}{2}, 1, \frac{M_s^2 b^2}{4}\right) \end{aligned} \tag{5.31}$$

When $\hat{s} > (M_P^{(n+2)} \pi^{n/2})/M_s^n \Gamma(n/2)$, $|\xi|$ is large compared to 1 and the exponentiation in Eq. 5.30 is important. This corresponds to the region of classical scattering.

In the opposite case, the importance of higher order corrections is negligible and the Born term is dominating. This corresponds to the region of quantum mechanical scattering.

5.3.4 Small Momentum Transfers: $\sqrt{\hat{s}} \ll M_s$

For small momentum transfers, i.e. $-\hat{t} \ll M_s^2$, t can be neglected in the nominator of Eq. 5.27:

$$A_{\text{Born}} = -\frac{\hat{s}^2}{M_{\text{eff}}^4}, \quad M_{\text{eff}} = \frac{1}{2}\left(\frac{(n-2)2^n \pi^{\frac{n-2}{2}} M_P^{n+2}}{M_s^{n-2}}\right)^{\frac{1}{4}}, \tag{5.32}$$

meaning that scattering is almost isotropic. KK propagators can be replaced by vertex factors so that the exchange of KK modes corresponds to a contact interaction. A geometric series of ladder diagrams is obtained and summed over so that unitarity is constrained:

$$A_{\text{ladders}} = \frac{A_{\text{Born}}}{1 - A_{\text{Born}} X}, \tag{5.33}$$

with $X \approx \frac{1}{32\pi^2}(\ln \frac{M_s^2}{\hat{s}/4} + i\pi)$. The Born term—neglecting spins of the colliding partons—can be written in terms of an effective Planck mass M_{eff}:

$$A_{\text{Born}} = -\frac{\hat{s}^2}{M_{\text{eff}}^4} \tag{5.34}$$

with:

$$M_{\text{eff}} = \frac{1}{2}\left(\frac{(n-2)2^n \pi^{\frac{n-2}{2}} M_P^{n+2}}{M_s^{n-2}}\right)^{\frac{1}{4}} \tag{5.35}$$

When $|A_{\text{Born}} X| \ll 1$, the Born term dominates. In that case, the cross section is written as

$$\frac{\mathrm{d}\sigma}{\mathrm{d}\hat{t}} = \frac{k_s}{\hat{s}}\left[\frac{\pi\alpha_s^2}{\hat{s}} f(z) - \frac{\hat{s}\alpha_s}{M_{\text{eff}}^4} g(z) + \frac{\hat{s}^3}{\pi M_{\text{eff}}^3} h(z)\right] \tag{5.36}$$

In the above equation, $z = \cos(\hat{\theta})$, $\hat{\theta}$ being the scattering angle in the center of mass system. The first term covers QCD processes, while the last term covers the gravitational part. The term in the middle is the interference between QCD and gravitational scattering. The functions $f(z)$, $g(z)$ and $h(z)$ and the constant k_s are process dependent and take into account the spin of the colliding partons. The definitions can be found in Ref. [12].

In the case of small momentum transfers, s-,t- and u-channels are equally efficient. The s-channel ladder diagrams are unitarized in a similar way. KK modes can also be produced on shell in this channel, but these stable gravitons would not be observable by a general purpose detector such as ATLAS (see Sect. 6.1) at the LHC.

The relevance of the u-channel contribution is suppressed for proton-proton collisions because of the low probability for a collision between two partons with identical flavor, spin and color.

Note that gravitational scattering will become weaker for smaller values of M_s (i.e. wider membranes), because the KK-modes with a mass above M_s are suppressed.

5.3.5 Experimental Limits

Experiments at the Tevatron have looked for gravitational effects, but have—so far—only been able to put limits on the model parameters. The most stringent limits come from dijet angular distribution measurements done by the D0 experiment [13]. Two different formalisms were investigated, namely the GRW model [10] and the HLZ model [9]. The model parameter is the effective Planck scale, M_S (not to be confused with the KK mas cut-off M_s), which is related to the fundamental Planck scale M_P used in this text in the following way:

$$M_P = 2^{\frac{1-n}{2+n}} \pi^{\frac{-n}{4+2n}} \Gamma \left(\frac{n}{2}\right)^{\frac{-1}{2+n}} M_S \tag{5.37}$$

The HLZ formalism also includes the subleading dependence on the number n of extra dimensions. None of these models contains a parameter comparable to M_s, i.e. the KK mass cut-off, discussed in this text; the KK mass cut-off was in both models fixed at the Planck scale, i.e. $M_s = M_S$. For six extra dimensions, a 95% confidence limit was observed at $M_P = 0.54$ TeV, with $M_s = 1.39$ TeV. For four extra dimensions the limit was set at $M_P = 0.80$ TeV, with $M_s = 1.66$ TeV.

5.4 Black Holes in the ADD Model

An exciting consequence of the ADD model is the possibility of the production of black holes (BHs) at the LHC and beyond. Black holes are well understood general-relativistic objects when their mass M_{BH} far exceeds the fundamental Planck mass M_P. But when their mass approaches M_P, the BHs become stringy and very hard to predict. In this section we will discuss the properties of light BHs by simple semiclassical arguments, which are strictly valid for $M_{BH} \gg M_P$. Because of the unknown stringy corrections, this is merely an approximation.

5.4.1 Production

The Schwarzschild radius r_s of a $(4+n)$-dimensional black hole is given by [5]:

$$r_s = \frac{1}{\sqrt{\pi} M_P} \left[\frac{M_{BH}}{M_P} \frac{8\Gamma(\frac{n+3}{2})}{n+2} \right]^{\frac{1}{n+1}} \tag{5.38}$$

Consider two colliding partons with $\sqrt{\hat{s}} = M_{BH}$. A black hole can be formed when the interacting partons come closer than twice r_s. This means that in the rest frame of the incoming partons their longitudinal wavelength $\lambda_l \propto 2/\sqrt{\hat{s}}$ and transverse wavelength $\lambda_T \propto 1/p_T$ need to be smaller than r_s.

This implies a minimum on the black hole mass, which lies in the neighborhood of the Planck scale [14]:

$$M_{\min,1} = M_P \left(\frac{(2\sqrt{\pi})^{n+1}(n+2)}{8\Gamma(\frac{n+3}{2})} \right)^{\frac{1}{n+2}} \tag{5.39}$$

The parton cross section for black hole production can be estimated from geometrical arguments and is of the order $\sigma(M_{BH}) \approx \pi r_s^2$. To calculate the production in hadron collisions, the convolution with PDFs needs to be made. It was calculated [8] that for a fundamental Planck scale of around 1 TeV, LHC will produce over 10^7 black holes a year for a luminosity of 30 fb^{-1}/year.

Another limit on the black hole's mass comes from the existence of the finite width of the membrane; in the approximation of a narrow width, the membrane cannot be more extended than the black hole itself. This means that the Schwarzschild radius r_s should not be smaller than $1/M_s$, which is only true for masses above $M_{\min,2}$:

$$M_{\min,2} = \frac{M_P^{n+2}(2+n)\pi^{\frac{n+1}{2}}}{8\Gamma(\frac{3+n}{2})M_s^{n+1}} \tag{5.40}$$

The maximum of Eqs. 5.40 and 5.39 is the minimum black hole mass possibly created. Table 5.1 shows the minimum mass for a few choices of M_P, M_s and n. For n=4 extra dimensions and for $M_s/M_P > 1.12$ ($M_s/M_P > 1.05$ for n=6), the minimum black hole mass equals Eqs. 5.39 and depends on M_P only. In the complementary region the minimum black hole mass is described by Eq. 5.40 and goes as $M_P/(M_s/M_P)^5$ ($M_P/(M_s/M_P)^7$ for n=6). In the latter case, small values of M_s, i.e. larger values of the membrane width, will prevent black holes from being created.

5.4.2 Decay

The decay of the BH is thermal and governed by its Hawking temperature T_H, given by

Table 5.1 Table showing for a few choices of M_P, M_s and n, the minimum black hole mass possibly created, defined as the maximum of Eqs. 5.39 and 5.40

M_P (TeV)	M_s (TeV)	n	$M_{\min}$ (TeV)
1	1	6	2.4
1	1	4	2.2
1	0.5	6	14.0
1	2	6	2.2
2	2	6	4.7
5	5	6	11.8

$$T_H = \frac{n+1}{4\pi r_s} \tag{5.41}$$

The above formula means that heavier BHs have colder decay products. To a first approximation, a BH is a point radiator and the average multiplicity $\langle N \rangle$ is given by:

$$\langle N \rangle \approx \frac{M_{BH}}{2T_H} \tag{5.42}$$

Equation 5.42 is only reliable when the mass of the BH is much larger than the Hawking temperature, i.e. $\langle N \rangle \gg 1$. The decay kinematics truncates the spectrum at energies $E \approx M_{MB}/2$.

The thermal decay of the BH means that the decay is flavor independent, with roughly equal probability to all particles of the SM (about 60). It is expected that about 10% of the particles are leptons and that only ~5% are neutrinos.

5.5 The GravADD Generator

5.5.1 Introduction

The GravADD generator [7] is a generator for gravitational scattering (GS) and black hole production (BH) in large extra dimensions, based on the theory discussed in the previous sections. Studies [7, 10, 15] have demonstrated that the Large Hadron Collider (LHC) is sensitive to these gravity mediated processes for certain parameter ranges. Given this motivation, the GravADD generator has been interfaced with the ATLAS software framework Athena [16], so that a comparison with ATLAS data becomes possible.

Because the standalone generator is not publicly available, we will only focus on its implementation in Athena. Details about the implementation and use from inside Athena were documented in an ATLAS internal note [17], and we will discuss the essential points in Sect. 5.5.4.

5.5.2 Monte Carlo Generators in Athena

The ATLAS software framework is called Athena and is written in C++ using Python for job configuration (so-called jobOptions files). Since most Monte Carlo generators are developed independently of the experiment, their libraries are kept as external libraries. Athena provides access to these libraries through so-called interface packages that allow for specifying parameters and calling the generator. The interface packages contain three major functions:

- initialize(): configuration of the generator.
- execute(): call the generator routines to generate the event, convert the event to HepMC format [18] and write the event to the transient store named StoreGate. The events can be read out of the StoreGate, converted to a ROOT-style format [19] and written out to a persistent store, namely an output file.
- finalize(): call destructors.

The GravADD generator uses mainly the PYTHIA source code, but contains a number of modified PYTHIA routines, as well as an external generator that is called by PYTHIA to generate the black hole production.

In the Athena implementation, GravADD is linked to the external PYTHIA library, while the modified functions and routines and the black hole generator are kept inside Athena, together with the interface class that steers GravADD.

5.5.3 Model Parameters

The parameters that describe the model are:

- M_P, the fundamental Planck Scale defined as in Eq. 5.18. Equation 5.18 is the default definition in GravADD, but it is also possible (see Sect. 5.5.4) for the user to change to the definition used by the Particle Data Group (Eq. 5.19) [6].
- M_s, the Kaluza Klein mass cut-off, with $1/M_s$ a measure of the width of the four-dimensional membrane.
- n, the number of extra dimensions, with $(n + 4)$ the total number of dimensions.
- M_{eff}, the effective Planck Scale which is used for the calculation of the gravitational scattering amplitude, see Eq. 5.32.

A set of parameters is given by M_{eff}, M_s/M_P (i.e. the relative KK mass cut-off) and the total number of dimensions.

5.5.4 Implementation in the ATLAS Software Framework Athena

The GravADD generator has been implemented in Athena and has become available from release 15.6 onwards. The code is available on the web.[2]

GravADD generates three different types of events: QCD, GS and BH. The generator assigns a flag to each event, gravev, to indicate if the event is a QCD (gravev = 0), a GS (gravev = 1) or a BH (gravev = 2) event. In the Athena implementation, this flag is stored in the StoreGate's weight container, allowing the user to retrieve this information afterwards in the analysis stage.

Gravitational scattering is implemented in PYTHIA 6.410 [20] by means of actual changes in the source code (see details later), while black hole production is plugged in as an external process using the UPINIT and UPEVNT routines that come with PYTHIA, and are called after initialization with PYINIT USER. The QCD processes are selected in the normal way, using the MSUB array in PYTHIA.

In order to maintain the distinction between external and internal code in the Athena implementation, the modified PYTHIA library is kept separated from the external black hole code by using two different packages; the package Generators/GravADD_i has been created to contain the code for the black hole production, while the modified PYTHIA library, called PythiaGS, has been added to the already existing Generators/PythiaExo_i package. We will give more details in the next paragraphs.

As mentioned earlier, the PYTHIA 6.410 library has been modified in order to incorporate the gravitational scattering. We refer to the PYTHIA manual [20] for more information about the standalone PYTHIA 6.410 program, and to the documentation inside the Generators/Pythia_i package for the implementation of PYTHIA 6.410 in Athena.[3]

The PythiaGS library distinguishes internal events (QCD and GS) from external ones (BH). For the generation of an event, instead of the ordinary PYEVNT call, PythiaGS calls its own subroutine PYEFFEVT, which calculates the probability that for an internal event, the event is not a QCD event, but a gravitational scattering one. In case of the latter, the program performs a proper color reconnection for the colorless graviton. In case of a QCD event, PYEFFEVT does the same as the ordinary PYEVNT routine in PYTHIA. Interference between QCD and gravitational scattering has been taken into account.

For the generation of an external event (BH), PythiaGS calls the black hole routines which are localized in the Generators/GravADD_i package. This package does not only contain the code for black hole generation, but also contains the interface code for reading in model parameters and other options from the Athena jobOptions file.

The black hole generator is a modified version of CHARYBDIS.1003 [21] which takes into account the extra requirement on the minimum black hole mass coming from the finite width of the membrane (see Eq. 5.40). The momentum scale for

---

[2] Available from http://alxr.usatlas.bnl.gov/lxr/source/atlas.

[3] Available from http://alxr.usatlas.bnl.gov/lxr/source/atlas/Generators/Pythia_i/doc/.

evaluating the PDFs is the inverse Schwarzschild radius rather than the black hole mass. The black hole is set to decay into all quarks, as well as into the W and the Higgs boson, with the particle types and energies chosen according to the gray-body modified emission probabilities and spectra. During the decay, the temperature of the black hole is allowed to vary. The termination of the decay occurs when the chosen energy for the emitted particle is ruled out by the kinematics of a two-body decay. At this point an isotropic 2-body decay is performed on the black hole remnant.

In order to be able to access the high-p_T phase space, GravADD performs a pre-weighted event generation. For GS and QCD events, the weight WTXS is assigned according to:

$$\mathsf{WTXS} = \left(\frac{p_T^2}{\frac{s}{800.0}} \right)^3 \tag{5.43}$$

In the PythiaGS code, this is done by modifying the ordinary PYTHIA PYEVWT routine, and by setting MSTP(142) = 1. In the next step, the event weight pari(10) is re-weighted with a factor $\sigma_{\mathrm{GS+QCD}}/\sigma_{\mathrm{QCD}}$ to take into account the GS.

In case of a black hole event, the pre-weighting of pari(10) is done by the external black hole code, and the weight is set to

$$\mathsf{WTXS} = \frac{0.5}{f_i(x_1, \mu_F^2),\, f_j(x_2, \mu_F^2)}, \tag{5.44}$$

with $f_i(x_1, \mu_F^2)$ and $f_j(x_2, \mu_F^2)$ the quark or gluon PDFs, defined at a factorization scale μ_F. Since the PDFs fall off with increasing x, the weight defined in Eq. 5.44 will favor high mass BH production.

Since black holes are included as external process, it is possible to switch off black hole production, see Sect. 5.5.5.

5.5.5 Use From Inside Athena

The configuration of GravADD is done through the so-called Athena jobOptions files, which are Python files where the user can set options for the generator [16]. The package Generators/GravADD_i contains the code to handle them.

Apart from the model parameters listed in Sect. 5.5.1, the user can also use the jobOptions service for setting the collision energy, turning off BH generation, and for changing the default definition of the fundamental Planck scale (see Eq. 5.18) to the definition used by the Particle Data Group [6] (see Eq. 5.19).

A detailed description of how to compose a jobOptions file and an example of use are given in Ref. [17].

Note that since GravADD generates weighted events, the user should always fill analysis histograms with the event weight pari(10), which is stored in the weight container. At the end of the event loop, the histograms need to be divided by the bin width and the total sum of the weights, and multiplied with the total integrated cross section, in order to convert the results to nb/(dimension of the horizontal axis). The total integrated cross section of event samples generated in Athena, is stored as MetaData in the logfile.

References

1. N. Arkani-Hamed, S. Dimopoulos, G. Dvali, The hierarchy problem and new dimensions at a millimeter. Phys. Lett. B **429**, 263 (1998) hep-ph/9803315
2. N. Arkani-Hamed, S. Dimopoulos, G. Dvali, Phenomenology, astrophysics and cosmology of theories with sub-millimeter dimensions and TeV scale quantum gravity. Phys. Rev. D **59**, 086004 (1999) hep-ph/9807344
3. M.J. Duff, Kaluza-Klein theory in perspective. pp. 22–35 (1995) hep-th/9410046
4. D.J. Kapner, T.S. Cook, E.G. Adelberger, J.H. Gundlach, B.R. Heckel, C.D. Hoyle, H.E. Swanson, Tests of the gravitational inverse-square law below the dark-energy length scale. Phys. Rev. Lett. **98**, 021101 (2007) hep-ph/0611184
5. R.C. Myers, M.J. Perry, Black holes in higher dimensional space-times. Ann. Phys. **172**, 304–347 (1986)
6. C. Amsler et al., Review of particle physics. Phys. Lett. B **667**, 1 (2008)
7. L. Lönnblad, M. Sjödahl, Classical and non-classical ADD-phenomenology with high-ET jet observables at collider experiments. JHEP **0610**, 088 (2006) hep-ph/0608210
8. S. Dimopoulos, G. Landsberg, Black holes at the LHC. Phys. Rev. Lett. **87**, 161602 (2001) hep-ph/0106295
9. T. Han, J.D. Lykken, R.-J. Zhang, On Kaluza-Klein states from large extra dimensions. Phys. Rev. D **59**, 105006 (1999) hep-ph/9811350
10. G.F. Giudice, R. Rattazzi, J.D. Wells, Quantum gravity and extra dimensions at high-energy colliders. Nucl. Phys. B **544**, 3–38 (1999) hep-ph/9811291
11. M. Sjödahl, G. Gustafson, Gravitational scattering in the ADD-model at high and low energies. EPJ, C **53**, 109 (2008) hep-ph/0608080
12. D. Atwood, S. Bar-Shalom, A. Soni, Dijet production at hadron colliders in theories with large extra dimensions. Phys. Rev. D **62**, 056008 (2000) hep-ph/9911231
13. The D0 Collaboration, Measurement of dijet angular distributions at $\sqrt{s}$ = 1.96 TeV and searches for quark compositeness and extra spatial dimensions. Phys. Rev. Lett. 103:191803, hep-ex/09064819
14. L. Lönnblad, M. Sjödahl, T. Åkesson, QCD-suppression by black hole production at the LHC. JHEP **0509**, 019 (2005) hep-ph/0505181
15. N. Boelaert , T. Åkesson, Dijet angular distributions at $\sqrt{s}$ =14 TeV. EPJ C **66**, 343–357 (2005) hep-ph/09053961
16. ATLAS Collaboration, ATLAS computing: technical design report. Technical design report ATLAS. CERN, Geneva (2005)
17. N. Boelaert, Implementation of the GravADD generator in Athena. Technical Report ATL-PHYS-INT-2010-012, CERN, Geneva (2010)
18. HepMC: a C++ Event Record for Monte Carlo Generators, http://http://savannah.cern.ch/projects/hepmc.
19. ROOT, A Data Analysis Framework, http://root.cern.ch.
20. T. Sjöstrand, S. Mrenna, P. Skands, PYTHIA 6.4 physics and manual. JHEP **0605**, 026 (2006) hep-ph/0603175
21. C.M. Harris, P. Richardson, B.R. Webber, CHARYBDIS: a black hole event generator. JHEP **0308**, 033 (2003) hep-ph/0307305

Chapter 6
The ATLAS Experiment

ATLAS is a general purpose detector, designed to investigate at high luminosity (up to $10^{34}\ \mathrm{cm}^{-2}\,\mathrm{s}^{-1}$)many different types of physics that might become detectable in the energetic collisions of the LHC. The ATLAS collaboration aims at performing high precision tests of QCD, electroweak interactions, and flavor physics. Furthermore, the search for the Higgs boson, incorporating many production and decay channels over a large mass range, is a very important aspect of the ATLAS physics program. Not only the Standard Model is subject of investigation, but also physics beyond the Standard Model forms an important area for study. Large extra dimensions, supersymmetry, quark compositeness and new heavy quarks are just a few topics that are grouped under this numerator. Signatures of new physics are quite often provided by high-p_T jet and missing transverse energy E_T^{miss} measurements. More information about the physics program can be found in Ref. [1].

6.1 Detector Layout

The overall detector layout of ATLAS is shown in Fig. 6.1. The detector is nominally forward-backward symmetric with respect to the interaction point and consists of a series of concentric detector systems around the interaction point. The main components of the detector are: the inner detector (ID) used for tracking of charged particles, the electromagnetic and hadronic calorimeters used for measuring the energy of electrons, photons and jets, the muon detector for identifying and measuring muons, and the magnet system. We will briefly discuss their main properties; more information can be found in Ref. [2, 3].

The coordinate system used to describe the ATLAS detector has the origin in the collision point, the z-axis parallel with the beam direction and the x-y plane transverse to it. The positive x-axis is defined as pointing from the interaction point to the center of the LHC ring and the positive y-axis is defined as pointing upwards. The azimuthal angle ϕ is measured around the beam axis, and the polar angle θ is measured from the beam axis. The pseudorapidity defined as $\eta = -\ln\tan(\theta/2)$ is

N. Boelaert, *Dijet Angular Distributions in Proton–Proton Collisions*, Springer Theses, DOI: 10.1007/978-3-642-24597-8_6, © Springer-Verlag Berlin Heidelberg 2012

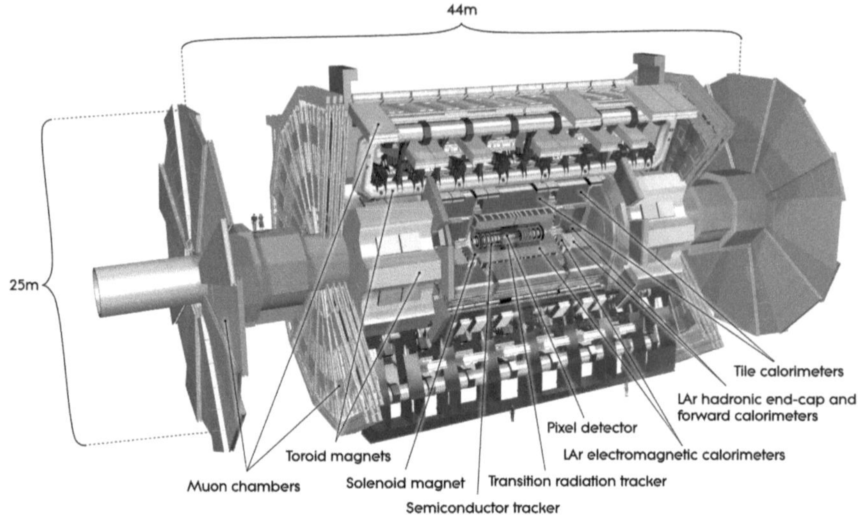

Fig. 6.1 Schematic view of the ATLAS detector

very often used instead of θ. Transverse quantities, such as transverse momentum p_T and transverse energy E_T are defined in the x-y plane.

A characteristic feature of the ATLAS detector is its enormous magnet system. The magnet configuration is based on an inner thin superconducting 2 T solenoid surrounding the inner detector cavity, and three large superconducting air-core toroids consisting of independent coils arranged with an eight-fold radial symmetry outside the calorimeters, see Fig. 6.2.

Inside the solenoid is the inner detector (ID), used for precise tracking of charged particles in the pseudorapidity range $-2.5 < \eta < 2.5$. The ID is contained within a cylinder of length 7 m and a radius of 1.15 m. Pattern recognition, momentum and vertex measurements, and electron identification are achieved with a combination of discrete high-resolution semiconductor pixel and microstrip (SemiConductor Tracker or SCT) detectors in the inner part of the tracking volume, and continuous straw-tube tracking detectors (Transition Radiation Tracker or TRT) with transition radiation capability in its outer part. Figure 6.3 gives a schematic view. The highest granularity is achieved around the vertex region using the pixel detector which consists of three cylindrical layers with radius 5, 8, 12 cm in the barrel region and two times three disks in the forward region. There are 1,744 modules, each module containing about 47,000 pixels which have an area of $50 \times 400\,\mu m^2$. The hit resolution is $15 \times 115\,\mu m$.

The SCT has four cylindrical double layers with radius 30, 37, 44, 51 cm and two times 9 disks in the forward region. There are 4,088 modules with 80 µm strips, using small (40 m rad) angle stereo to obtain the z measurement, allowing for a hit resolution of $17 \times 580\,\mu m$.

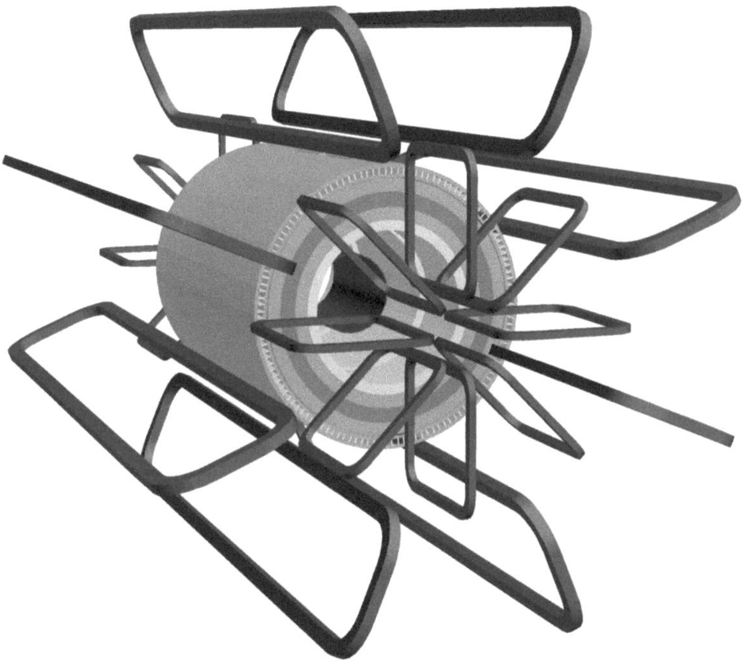

Fig. 6.2 Geometry of magnet windings and tile calorimeter steel. The eight barrel toroid coils, with the end-cap coils interleaved are visible. The solenoid winding lies inside the calorimeter volume. The tile calorimeter is modelled by four layers with different magnetic properties, plus an outside return yoke. For the sake of clarity the forward shielding disk is not displayed

The TRT is constructed from straw tubes with a diameter of 4 mm and with 35 μm anode wires. The straws are arranged in 73 layers in the barrel region (straws along beam-axis) and 2×160 layers (disks) in the end-cap region (straws radially placed). A hit resolution of 130 μm per straw is obtained.

The combination of precision trackers at small radii with the TRT at a larger radius gives very robust pattern recognition and high precision in both R-ϕ and z coordinates. The straw hits at the outer radius contribute significantly to the momentum measurement, since the lower precision per point compared to the silicon is compensated by the large number of measurements and longer measured track length.

A schematic view of the ATLAS electromagnetic and hadronic calorimeter system is given in Fig. 6.4, with its main parameters listed in Table 6.1. The calorimetry consists of an electromagnetic (EM) calorimeter covering the pseudorapidity region $|\eta| < 3.2$, a hadronic calorimeter covering $|\eta| < 3.2$ and forward calorimeters covering $3.1 < |\eta| < 4.9$.

The central part of the EM calorimeter covers pseudorapidity values up to $|\eta| < 3.2$ with an accordion geometry made of liquid argon and lead. The accordion structure ensures azimuthal uniformity (no cracks). Over the region devoted to precision physics, i.e. $|\eta| < 1.475$ (the Electromagnetic Barrel Calorimeter, EMB), the

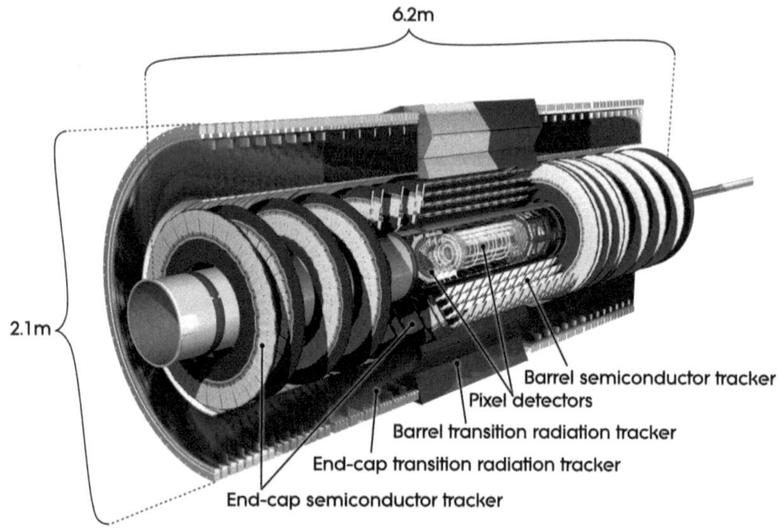

Fig. 6.3 Cut-away view of the ATLAS inner detector

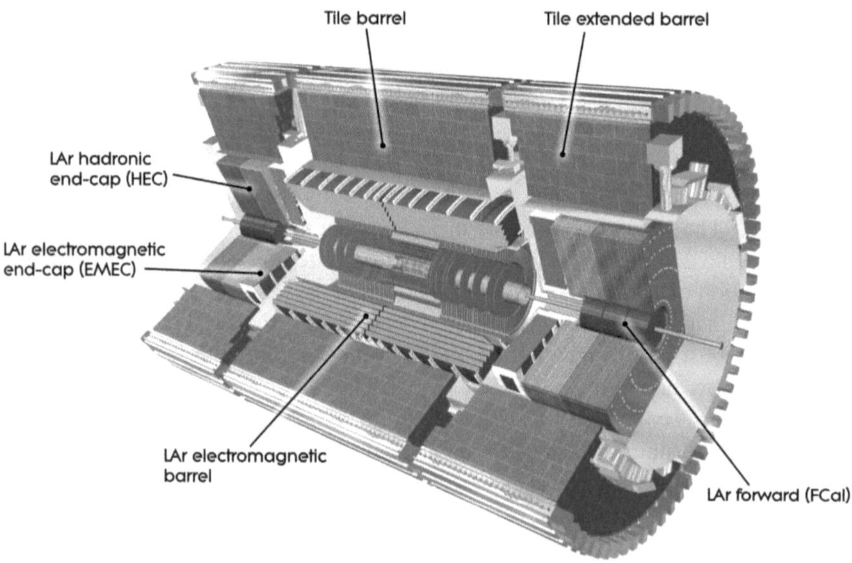

Fig. 6.4 Cut-away view of the ATLAS calorimeter system

Table 6.1 The main parameters of the ATLAS calorimeter system

EM calorimeter	Barrel (EMB)	End-cap EMEC
Coverage	$\|\eta\| < 1.475$	$1.375 < \|\eta\| < 3.2$
Longitudinal segmentation	3 samplings	2–3 samplings
Granularity ($\Delta\eta \times \Delta\phi$)		
Sampling 1	0.003×0.1	0.003×0.1
Sampling 2	0.025×0.025	0.025×0.025
Sampling 3	0.05×0.025	0.05×0.025
Number of channels	102,400	62,208
Presampler	*Barrel*	*End-cap*
Coverage	$\|\eta\| < 1.475$	$1.5 < \|\eta\| < 1.8$
Longitudinal segmentation	1 sampling	1 sampling
Granularity ($\Delta\eta \times \Delta\phi$)		
Sampling 1	0.025×0.1	0.025×0.1
Number of channels	7,808	768
TileCal	*Barrel*	*Extended barrel*
Coverage	$\|\eta\| < 1.0$	$0.8 < \|\eta\| < 1.7$
Longitudinal segmentation	3 samplings	3 samplings
Granularity ($\Delta\eta \times \Delta\phi$)		
Samplings 1 & 2	0.1×0.1	0.1×0.1
Sampling 3	0.2×0.1	0.2×0.1
Number of channels	5,760	1,792
Hadronic end-cap (HEC)		
Coverage	$1.5 < \|\eta\| < 3.2$	
Longitudinal segmentation	4 samplings	
Granularity ($\Delta\eta \times \Delta\phi$)		
$1.5 < \|\eta\| < 2.5$	0.1×0.1	
$2.5 < \|\eta\| < 3.2$	0.2×0.2	
Number of channels	3,072	
Forward calorimeters		
Coverage	$3.1 < \|\eta\| < 4.9$	
Longitudinal segmentation	1 EM sampling 2 Hadronic samplings	
Granularity ($\Delta\eta \times \Delta\phi$)		
Sampling 1	$\approx 0.1 \times 0.1$	
Sampling 2 & 3	$\approx 0.2 \times 0.2$	
Number of channels	1,792	

EM calorimeter is segmented in three sections in depth, with a fine granularity of typically $\Delta\eta \times \Delta\phi = 0.025 \times 0.025$. For the end-cap inner wheels (Electromagnetic End-cap Calorimeter, EMEC), $1.375 < |\eta| < 3.2$, the calorimeter is segmented in two sections in depth and has a coarser lateral granularity. The total thickness of the EM calorimeter is > 24 radiation lengths[1] in the barrel and > 26 in the end-caps.

A Presampler (PS) is placed in front of the electromagnetic calorimeters and extends from $-1.8 < \eta < 1.8$. Its purpose is to estimate the energy lost in material between the interaction point and the calorimeter.

The EM calorimeters have a total of about 180,000 cells. Ionization signals are summed and brought through feedthroughs to exit the cryostats. Next, they are amplified, shaped and digitized in Front End Boards (FEBs) which are located inside Front End Crates (FECs). A calibration board is placed in each FEC in order to generate and distribute adjustable currents to all cells readout through the feedthrough. The calibration board has 128 channels used for injecting charge.

The hadronic calorimeter is surrounding the EM calorimeter. The bulk $|\eta| < 1.7$ of the hadronic calorimetry is provided by a novel scintillator-tile calorimeter—and is therefore called TileCal—which is separated into a large barrel and two smaller extended barrel cylinders, one on each side of the barrel. Between the modules are 700 mm wide gaps that are needed for the services of the inner detector and the central EM calorimeter. The radial depth of the TileCal is approximately 7–8 pion interaction lengths[2]. All TileCal barrels are divided along into 64 modules in ϕ ($\Delta\phi \approx 0.1$), and three radial layers, containing cells with $\Delta\eta = 0.1$ for the first two layers and $\Delta\eta = 0.2$ for the last layer. TileCal is a sampling calorimeter using iron as absorber and tiles of scintillating plastic as the active material. The tiles are oriented perpendicularly to the beam direction and staggered in radial depth. The light readout is done by optical fibers, that couple to photomultiplier tubes in the outer part of each module. The outer part of each module also contains several detector electronics services: the electronics for the analog trigger, signal shaping, digitization and integration, the charge injection calibration system, the low voltage power supply and a set of optical fibers dedicated to the laser calibration system.

The two hadronic end-caps (HECs) of the ATLAS detector occupy the region from 4,200 to 6,118 mm measured axially from the interaction point and 1.5 to 3.2 in pseudorapidity, and overlap partially with the tile hadronic calorimeter and the forward calorimeters. The HEC is a liquid argon sampling calorimeter with copper absorber plates that are oriented parallel and perpendicular to the beam direction. Each end-cap consists of two separate wheels, formed from 32 pie-shaped modules, located directly behind the end-cap electromagnetic calorimeter and sharing the same liquid argon calorimeter cryostats.

The forward calorimeter (FCAL) system covers the region very close to the beam pipe, from $3.1 < |\eta| < 4.9$, and is therefore a particularly challenging detector owing to the high level of radiation it has to cope with. The FCAL is a liquid argon sampling calorimeter and consists of three sections: the first one is an electric layer with a copper absorber, while the other two are hadronic layers made out of tungsten

---

1 A radiation length is defined as the thickness of material crossing for which an electron has a probability $P = 1 - 1/e$ (or roughly 63%) of radiating a photon.

2 The interaction length of a particle is its mean free path before undergoing an interaction that is neither elastic nor quasi-elastic (diffractive).

absorbers. The FCAL modules are cylindrical in shape with a coaxial hole through which the LHC beams pass. Liquid argon is chosen because of its radiation hardness and because the HEC is made of liquid argon as well. The FCAL system is placed adjacent to the other calorimeters in the two end-cap cryostats, relatively close to the interaction point, providing a nearly seamless calorimetry and a natural shielding for the muon system.

The set of calorimeters made from liquid argon is often called the LAr.

The LAr calorimetry is contained in a cylinder with an outer radius of 2.25 m and extends longitudinally to ± 6.65 m along the beam axis. The outer radius of the scintillator-tile calorimeter is 4.25 m and its half length is 6.10 m. The total weight of the calorimeter system, including the solenoid flux-return iron yoke which is integrated into the tile calorimeter support structure, is about 4,000 Tons.

The readout of the calorimeters is highly granular for the electromagnetic devices, with typically three longitudinal segments with varying cell sizes, e.g. $\Delta\eta \times \Delta\phi = 0.025 \times 0.025$ in the second segment, containing the electromagnetic shower maximum. The hadronic calorimeters are coarser, with typically $\Delta\eta \times \Delta\phi = 0.1 \times 0.1$, but also have at least three shower segments. The total thickness of the ATLAS calorimeter system is at least 10 hadron absorption lengths over the whole acceptance region.

The calorimeter system is surrounded by the muon spectrometer (cfr. Fig. 6.5) which is based on the magnetic deflection of muon tracks in the large superconducting air-core toroid magnets. The magnet system, with a long barrel and two inserted end-cap magnets, generates a large magnetic field volume with strong bending power. The bending power can be characterized by the field integral $\int Bdl$, i.e. the integration over the magnetic field component orthogonal to the muon direction and along the trajectory of the muon. For the pseudorapidity range $0 < |\eta| < 1.4$, values of the field integral ranging from 1.5 to 5.5 Tm are obtained. The magnets have a light and open structure, minimizing multiple-scattering effects, and good muon momentum resolution is achieved with three stations of high-precision tracking chambers. In the barrel region, tracks are measured in chambers arranged in three cylindrical layers (stations) around the beam axis; in the transition and end-cap regions, the chambers are installed vertically, also in three stations. Over most of the pseudorapidity-range, a precision measurement of the track coordinates in the principal bending direction of the magnetic field is provided by Monitored Drift Tubes (MDTs). At large pseudorapidities and close to the interaction point, Cathode Strip Chambers (CSCs) with higher granularity are used in the innermost plane over $2 < |\eta| < 2.7$. For muons with an energy less than 1 TeV, a stand-alone momentum resolution better than 10% is obtained.

The muon spectrometer defines the overall dimensions of the ATLAS detector. The outer chambers of the barrel are at a radius of about 11 m. The half-length of the barrel toroid coils is 12.5 m, and the third layer of the forward muon chambers, mounted on the cavern wall, is located about 23 m from the interaction point. The overall weight of the ATLAS detector is about 7,000 Tons.

In order to perform luminosity measurements, ATLAS has a number of detectors placed in the forward region. LUCID, a LUminosity Cherenkov Integrating Detector,

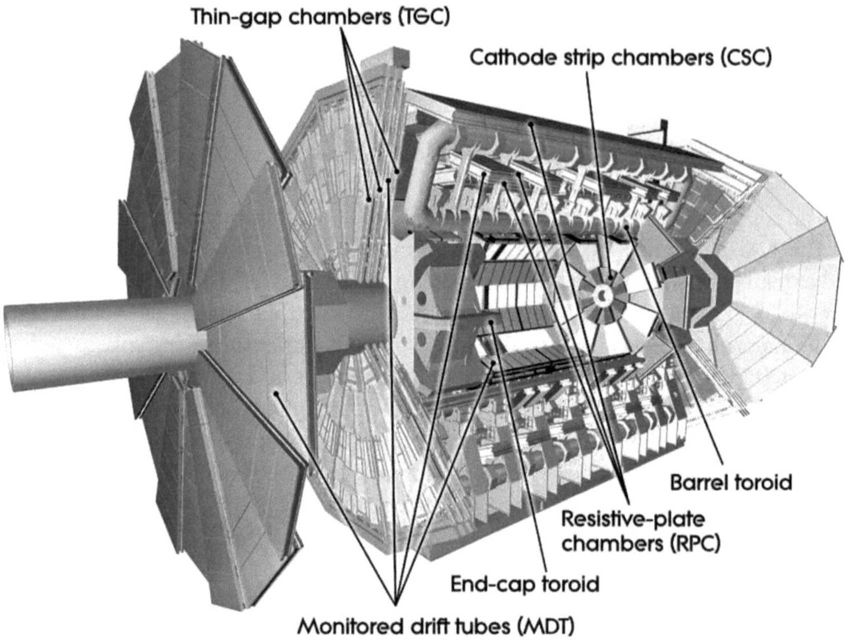

Fig. 6.5 Cut-away view of the ATLAS muon system

monitors the luminosity by counting the number of inelastic interactions for each bunch crossing in projective Cerenkov counters placed at 17 m from the center of ATLAS.

The Absolute Luminosity for ATLAS (ALFA) detector performs a luminosity calibration by measuring the elastic scattering cross section in the region of Coulomb Nuclear interference.

The ATLAS detector main performance goals are listed in Table 6.2. It is important to note that, for high-p_T muons, the muon-spectrometer performance as given in Table 6.2 is independent of the inner-detector system. Also note that a big fraction of the jet energy is measured in the electromagnetic calorimeters (see also the next chapter, more specifically Fig. 7.2).

6.2 Trigger

The mean number of interactions per bunch crossing at the design luminosity of $10^{34}\,\mathrm{cm}^{-2}\,\mathrm{s}^{-1}$ is 23, while the bunch spacing is 25 ns. This means that the proton-proton interaction rate is approximately 1 GHz, while the event data recording, based on technology and resource limitations, is limited to about 200 Hz. This requires an overall rejection factor of 5×10^6 against minimum-bias processes while maintaining

Table 6.2 General performance goals of the ATLAS detector [2]. The units for E and p_T are given in GeV

Detector component	Required resolution	η Coverage measurement	Trigger
Tracking	$\sigma_{p_T}/p_T = 0.05\% p_T \oplus 1\%$	± 2.5	
EM calorimetry	$\sigma_E/E = 10\%\sqrt{E} \oplus 0.7\%$	± 3.2	± 2.5
Hadronic calorimetry (jets)			
Barrel and end-cap	$\sigma_E/E = 50\%\sqrt{E} \oplus 3\%$	± 3.2	± 3.2
Forward	$\sigma_E/E = 100\%\sqrt{E} \oplus 10\%$	$3.1 < \lvert\eta\rvert < 4.9$	$3.1 < \lvert\eta\rvert < 4.9$
Muon spectrometer	$\sigma_{p_T}/p_T = 10\%$ at p_T = 1 TeV	± 2.7	± 2.4

maximum efficiency for the new physics. Therefore, a dedicated trigger system is necessary.

The ATLAS trigger system has three levels: L1, L2 and the event filter. L1 is a very fast hardware implemented trigger, but L2 and the event filter are software implemented triggers. The combination of L2 and event filter is called the high-level trigger.

Each trigger level refines the decisions made at the previous level and, where necessary, applies additional selection criteria. The data acquisition system receives and buffers the event data from the detector-specific readout electronics, at the L1 trigger accept rate. There are over 1,600 point-to-point readout links. The first level uses a limited amount of the total detector information to make a decision in less than 2.5 μs, reducing the rate to about 75 kHz. The two higher levels access more detector information for a final rate of up to 200 Hz with an event size of approximately 1.3 Mb.

6.2.1 The Jet Trigger Slice

The *jet slice* is the subset of the ATLAS trigger for jet selection. The jet slice at L1 works with a fast sliding window in (η, ϕ) of dimension 0.8×0.8 in order to find regions of local energy maxima, which is replaced by a simple cone algorithm at L2. The event filter is running the so called ATLAS Cone algorithm. Jet algorithms are subject of discussion in Sect. 7.2. A study dedicated to jet algorithms [4] has concluded that a more suitable algorithm for the event filter exists, namely the anti-k_T algorithm. However the ATLAS Cone will still be used for the early data taking period.

There are two broad classes of jet triggers: single jet triggers which are triggering on the highest jet in the event, and multijet triggers. The exact definition of the trigger (threshold, prescale, etc) depends on the LHC collision energy, the luminosity and the physics priority.

6.3 Event Reconstruction

The event filter assembles data into several physics streams and one express stream and sends them to the CERN Tier-0 center where they are stored on tape. The assignment of a given event to a certain physics stream (e.g. jet stream, minimum bias stream, muon stream, etc.) is based on the trigger signature. The express stream contains about 10% of the recorded data, mostly high-p_T events with leptons or jets. The physics streams are grouped together in what is called the "bulk" data. Apart from express and physics streams, there are also streams used for calibration purposes.

The processing of the express stream starts promptly (in quasi-real time) after the data are sent to the Tier-0. Also the calibration streams are processed in quasi-real time. But the bulk reconstruction cannot start before the express stream is fully reconstructed and an assignment of the quality of the data is made. If problems with the express stream occur (bad calibration, software problems ...), they need to be solved and the express stream needs to be processed again. A "sign-off" procedure using the results from the data quality (DQ) assessment of express and calibration stream monitoring, determines whether Tier-0 bulk data processing can begin. This procedure is based on automated checks on histograms representing the data quality. Detailed information can be found in the next section.

Once the green light has been given for the bulk processing, the Tier-0 will start to produce Event Summary Data (ESDs, created from the RAW data) and Analysis Object Data (AODs, reduced event representation derived from ESDs). Both ESDs and AODs are transferred to different computing facilities spread over the world, using the computing grid. These centers are so-called Tier-1 centers and apart from being a storage facility, they are also re-reprocessing the ESDs and AODs (e.g. with an updated calibration).

Derived physics datasets (DPDs, derived from AODs) produced by the physics groups are copied to the Tier-2 facilities for further analysis. The Tier-2 facilities also provide the simulation capacity for the experiment, with the simulated data housed at Tier-1s. In addition, Tier-2 centers will provide analysis facilities and some will provide the capacity to produce calibrations based on processing some raw data.

This way, the ATLAS Computing Model [5] embraces a high degree of decentralisation and sharing of computer resources, and enables for all members of the ATLAS Collaboration speedy access to all reconstructed data for analysis during the data-taking period, and appropriate access to raw data for organised monitoring, calibration and alignment activities.

6.4 Data Quality

As mentioned in the previous section, the ATLAS experiment continuously produces large streams of raw data that need to undergo various stages of processing. In order to be able to diagnose problems in the processing phase as soon as they occur—

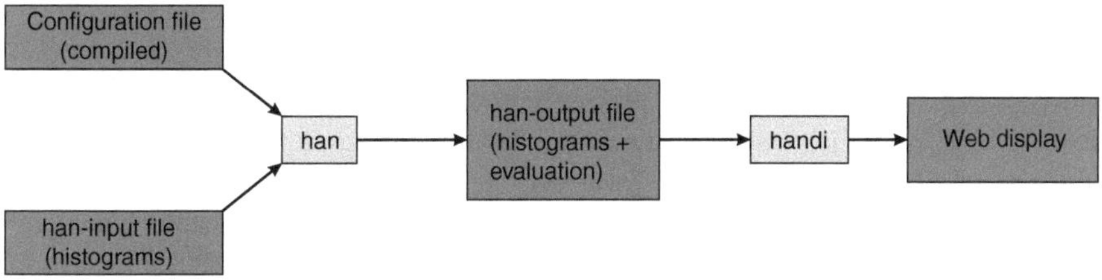

Fig. 6.6 Flowchart showing input and output of han and handi

so as to avoid processing delays—and to verify the data integrity, tools for prompt monitoring and archiving of the quality of the processed data have been developed and have become an essential part of the data processing chain. This collection of software is part of the ATLAS Data Quality Monitoring Framework (DQMF) [6], which embraces both online (before processing) and offline (after processing) data quality operations. The sections below will only focus on the quality of the *processed* data, since from an analysis point of view, this is the most important aspect of DQMF. As already mentioned at the beginning of this thesis, an ATLAS internal note was published in Ref. [7].

During the processing of the data, distributions representing the quality of the data are continuously being produced. The data quality distributions of one run (60 min) are merged in so-called monitoring ROOT files [8], containing histograms that either cover the whole run (providing *global run* information), or only part of it, of the order of a few minutes (providing *minutes block* information). These histograms serve as input for the evaluation of the quality of the run, which is done by running automated tests on them.

6.4.1 Automatic Evaluation and Display of Data-Quality Histograms: Han *and* Handi

The core utilities among those developed for data-quality assessment of processed data are the histogram analyzer, han, and the histogram-analysis display, handi. These utilities have been designed to be used in both standalone modes and as functions called from other applications. The han application performs an automatic evaluation of a set of histograms based on a user-defined configuration file, and the histograms and evaluation results are stored in a binary file. The handi application takes the han-output file as input and creates a tree of HTML files for viewing the histograms and associated evaluations. Figure 6.6 shows this schematically.

6.4.1.1 The Histogram Analyzer: Han

The han application is designed to promptly perform automated histogram checking using ROOT. Each distribution is checked according to a predefined algorithm and is assigned a DQ status flag; "green" for good, "yellow" for flawed (questionable, use with care but likely recoverable), "red" for bad, "gray" for undefined and "black" for disabled. han is implemented in C++ classes using libraries from ROOT and from the ATLAS Data Quality Monitoring Framework. The primary design considerations for this program are reliability and flexibility to be used as a component with other software.

The program is given two input files, one with the histogrammed data to be analyzed and one with a compiled configuration, and it produces one binary output file with the results of the analysis (see Fig. 6.6). The analysis takes a stepwise approach; first it checks individual histograms and assigns them a status, then it combines histograms into groups and provides summary information for those, based on the status of the individual histograms they contain. Finally it combines different groups and produces a summary based on the status of the subgroups. This hierarchical structure allows the result of a single histogram to propagate up to the highest level. This way, the user can easily detect data quality problems from the top-level by following the tree structure down to the level where they occur.

The han configuration format allows for the specification of the histograms to check, the algorithms to check them with, thresholds for yellow and red flags, and the place of each checked histogram in the data quality hierarchy. Furthermore, options for displaying the results with handi need to be set in the han configuration as well.

As a simple example of how this program is intended to work, suppose one has a set of histograms of a quantity that is distributed as a Gaussian. To quickly determine if the widths of all the distributions are as expected, one would provide these histograms as input to han, along with a configuration that specifies an algorithm that evaluates widths. The output would then have the information to quickly alert the user to unexpected distributions, along with the relevant information about how those decisions were reached.

The distributions, together with their status and summary information about the algorithm, are saved in a new ROOT file, which can be opened for inspection in ROOT. But in order to provide a more user friendly interface and a world wide accessibility of the data, handi—the han display—was developed, which translates the han-output into a HTML coded web display.

6.4.1.2 The Histogram-Analysis Display: Handi

handi is written in Python and uses, apart from standard Python modules, a C++ library of relatively simple, statically linked functions designed for reading and processing the han results.

The han-output file contains T Directories with groups of histograms that are structured in a directory tree defined by the user through the han configuration input

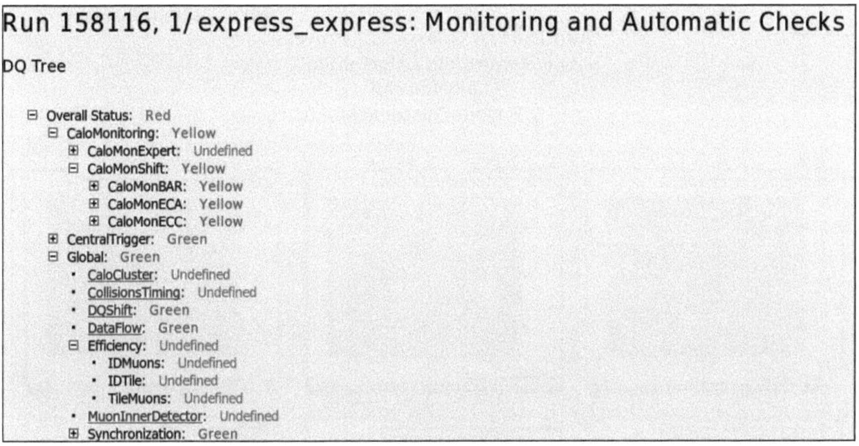

Fig. 6.7 Screenshot of the output of handi at the *top* level

file. Apart from the histogram itself, information about the algorithm configuration parameters and the results of the histogram check are stored in the output as well. Not only the histograms, but also every TDirectory is assigned a status—defined by the histograms it contains—by han, and all relevant information about this process is saved in the output file as well.

The goal of handi is to display the histograms and all information, maintaining the han-output directory tree structure. World wide accessibility is guaranteed by making a HTML web browser display.

handi is designed in a way so that the HTML pages offer information at three levels, going from very general to histogram specific. Figures 6.7 and 6.8 shows snapshots of these three levels.

First there is the **top-level page** (Fig. 6.7), which is an overview page displaying the complete directory structure of the han-output file, listing the name of the directory together with its status (which is colored accordingly).

When a certain directory contains histograms or subdirectories, the user can click on the name in order to get linked to a **directory page** (Fig. 6.8 top) which shows thumbnails of all histograms and subdirectories in that directory. The thumbnails are displayed in a colored frame according to their status.

In order to view detailed information about the histogram, the user can click on the thumbnail to get linked to the **one-histo page** (Fig. 6.8 bottom), containing an enlarged view of the histogram, together with extra information such as the algorithm configuration parameters (which algorithm, what thresholds) and the evaluation of the check (status and algorithm specific info). Basically all information that is found in the han-output file is displayed.

The histograms are loaded from the han-output file, plotted on ROOT canvas and saved in png format. Whenever an algorithm uses a reference histogram, the

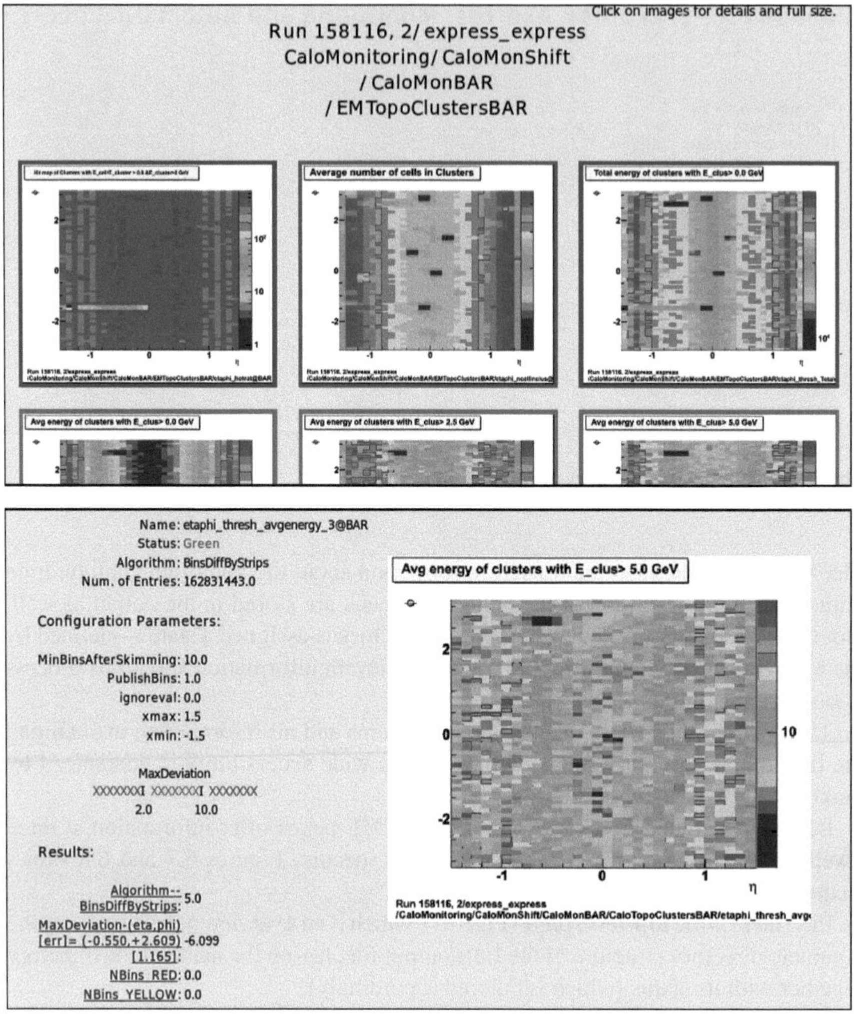

Fig. 6.8 Screenshots of the output of handi. Information is provided at three levels: *top*-level (see Fig. 6.7), directory level (*top* plot) and one-histo level (*bottom* plot)

reference is shown as well. Through the han configuration file, the user can give in options related to the plotting style (logarithmic axis, axis range, etc.).

Python modules are used to automatically generate the html files.

As for the input of handi, apart from the han-output file, the user also has to give in a title that will be used on every page and the location of where the HTML files need to be stored. There are also a few extra parameters that are used internally, such as the location of javascripts that run on the pages.

6.4.2 Prompt Assessment of Data-Quality During Data Taking

Use of the core utilities during the processing of recorded data requires a set of specific tools appropriate to the running environment. The data are processed in parallel on $\mathcal{O}(1000)$ machines, producing output files that must be combined and organized by data type and time interval. Files are merged so that the contents of the histograms and other plots are combined appropriately, in a configurable manner. As merged files are produced and updated while a run is being processed, they are displayed on the web. After a run has been completely processed, the final histograms are displayed, and the automatic assessments are uploaded to a database.

Within the ATLAS software environment, the AthenaMonitoring package provides common functionality for histogramming quantities during event reconstruction. A crucial feature of this package is that users define logical histogram locations within an output file; the physical placement of the histogram within the output file is determined by the framework in AthenaMonitoring. In this way, histograms can be placed in a directory structure ensuring that the proper time-granularity is attributed. For example, the top-level directory may be labelled according to the run number, and subdirectories may correspond to sub-run intervals within that run. Also, additional information may be saved along with the histograms for later bookkeeping or other operations. These metadata may be derived from the information provided when the histograms are registered with the application. An important use of the metadata is to specify which algorithm is used to combine instances of a particular histogram in the file-merge process.

Additionally, the AthenaMonManager holds general information about the running environment, and it makes this information available to the monitoring tools it manages. In this way, users can define and configure histograms according to information that is only available at run time. Also, different sets of histograms can be produced for different stages of event reconstruction; this is used to limit the number of histograms produced during memory-intensive processes by producing only those that cannot be made at a later stage.

Because the event-reconstruction programs run on many processors in parallel, many histogram files are produced with histograms filled with data from only a part of a run. To evaluate complete time intervals, all the histograms with partial statistics are combined into final histograms with full statistics. The application that merges the histogram files in this manner is DQHistogramMerge.

The DQHistogramMerge is executed at regular time intervals while a run is being processed. For input, it is given all the histogram files produced since the last execution. This way, the final merged file for a run is built in chunks. Since the raw data files are not necessarily processed in the order in which they were produced, it is crucial that the histograms are organized in a directory structure corresponding to sub-run time intervals. Since the directory structure is maintained during the merge process, all of the merged files, partial and final, correctly attribute time intervals to histograms regardless of the order in which the data are reconstructed.

The application DQWebDisplay is the heart of the web-display production that runs at Tier-0. It ties the core utilities han and handi together with scripts to upload data to the web server, to automatically generate index pages, and to upload results to a database.

The configuration of the web display contains a number of settings, but these settings are all correlated with the type of run that is being processed. To enhance reliability and reproducibility, the complete configurations are encoded in Python classes that are fixed for a given release of the ATLAS software. The selection of a particular configuration at runtime is done with a single string provided to DQWebDisplay.

The execution of DQWebDisplay is as follows. First, the input histogram file is analyzed by han; each distinct time interval contained in the input file is analyzed separately, so that after all analyses are complete, there are separate results files for these time intervals, e.g. a run file, several 10 min files, and several 30 min files. Second, handi is run on each of the results files, producing separate hierarchies of web pages for each time interval. Third, an index web page is generated for the run, with links to each of the separate hierarchies. Fourth, all of the HTML and image files are transferred to the web server. Finally, a script is executed on the web server that builds a list of all the runs that are present and generates a central web-page index. The final step ensures that the index page is always available and current, even if some other part of the script fails.

The DQWebDisplay application can be run in a special mode for producing web pages while a run is still being processed. In this case, it is given as input a histogram file with partial statistics, produced by one of the periodic executions of DQHistogramMerge. When running in this mode, DQWebDisplay maintains a local merged file with all of the statistics seen up that point: it runs DQHistogramMerge on the input file and the existing local merged file to produce a new local file. This procedure is not completely reliable, in the sense that the local file is not archived and not generally reproducible. However, since it operates entirely outside of the production system, it prevents the production system itself from being disrupted by the additional complexity of this process. These partial-statistics web pages are deleted when the full run has been processed and the final web pages are produced.

Apart from automated checks and a web display, the DQMF also provides a tool for monitoring the history, i.e. the evolution over several runs, of the DQ results. Furthermore, the most relevant results of the automated histogram checks, i.e. the status flags green, yellow, red, gray and black, are archived in the ATLAS condition database [5].

The archived DQ status flags are then used to filter on good data runs and luminosity blocks (corresponding to approximately two minutes of data taking with a constant instantaneous luminosity), by appliying DQ flag selection criteria. The resulting lists are also known as good run lists (GRLs). We will discuss the good run list criteria for studies with jets and missing transverse energy in Chap. 8.

References

1. ATLAS Collaboration, ATLAS detector and physics performance: technical design report. Technical report, CERN, Geneva (1999)
2. ATLAS Collaboration, The ATLAS Experiment at the CERN Large Hadron Collider. J. Instrum. **3**, S08003 (2008)
3. ATLAS Collaboration, Expected performance of the ATLAS experiment: detector, trigger and physics. CERN, Geneva (2008)
4. L. Asquith et al., Technical report ATL-COM-PHYS-2009-630. CERN, Geneva, Dec (2009)
5. D. Adams et al., The ATLAS computing model. Technical Report ATL-SOFT-2004-007, ATL-COM-SOFT-2004-009, CERN-ATL-COM-SOFT-2004-009, CERN-LHCC-2004-037-G-085, CERN, Geneva (2004)
6. N. Corso-Radu, H. Hadavand, M. Hauschild, R. Kehoe, S. Kolos, Data quality monitoring framework for the ATLAS experiment at the LHC. IEEE Trans. Nucl. Sci. **55**, 417–420 (2008)
7. N. Boelaert et al., Software design for prompt assessment of time-varying data quality. Technical Report ATL-COM-GEN-2010-002, CERN, Geneva (2010)
8. ROOT, A data analysis framework. http://root.cern.ch

Chapter 7
Jet Reconstruction

7.1 General Approach

A jet is a narrow cone of hadrons and other particles. Jets have to be defined by an algorithm and the definition of a jet is therefore unavoidably ambiguous; many jet finding algorithms exist and the best choice often depends on the topology of interest.

A jet algorithm starts from a list of input particles and tries to associate them to jets, so that the kinematic properties of the jets (e.g. momenta) can be related to the corresponding properties of the energetic partons produced in the hard scattering process. The input particles can be partons or hadrons in a theoretical calculation, but they can also be calorimeter signals at the experimental level. Quite often, particles that do not interact with material (e.g. neutrinos) or parent particles that decay too quickly to interact with the detector (lifetime less than 10 ps), are disregarded (although the decay products may be included). When comparing experiment with theory, it is important to use the same jet algorithm to ensure consistency.

A jet algorithm has two main elements; there is the *jet finding algorithm*, grouping together the particles that belong to a common jet, and there is the *recombination scheme*, defining how to add the momenta of the particles when combining them.

ATLAS has many different jet finding algorithms, and we will discuss them in the next section (Sect. 7.2), but the recombination scheme is always four-momentum addition. Section 7.3 specifies the input for ATLAS jet finders.

Apart from the ambiguity intrinsic to the jet algorithm, a further experimental complication to jet reconstruction is the jet energy calibration. A jet measurement in the ATLAS detector starts from signals recorded in the calorimeter cells which have been calibrated at the *electromagnetic* (EM) scale. This scale is set in test beams and/or in situ measurements and is defined to reproduce correctly the response to electromagnetic showers. But the electromagnetic scale does not reproduce the energy of hadrons correctly because hadrons deposit less energy in the calorimeters. Typically, the ATLAS calorimeter response to electrons is about a factor 1.3 higher than the hadron response. This is due to energy losses in forms not measurable, such as nuclear break-up, spallation, and excitation, energy deposits arriving out of

N. Boelaert, *Dijet Angular Distributions in Proton–Proton Collisions*, Springer Theses,
DOI: 10.1007/978-3-642-24597-8_7, © Springer-Verlag Berlin Heidelberg 2012

the sensitive time window (such as delayed photons), soft neutrons, and particles escaping the detector (neutrinos). Furthermore, the ATLAS magnetic field is able to bend low energetic charged particles out of the jet cone, and part of the energy is lost because of cracks, dead material and gaps in the calorimeter. Therefore a hadronic calibration is necessary in order to get to the *hadronic* level, i.e. the scale at which hadrons are defined. All effects from the detector itself are removed in this step. ATLAS has opted for a calibration style following a factorized approach, ensuring flexibility, which is desired in the early stage of ATLAS data taking where experience still needs to be gained. We will discuss several hadronic calibration methods in Sect. 7.4. A more detailed description can be found in Ref. [1].

7.1.1 Jet Reconstruction Performance Studies

Using a combination of Monte Carlo and data-driven calibration steps, ATLAS aims at obtaining an initial uncertainty of the jet energy scale between 5 and 10%.

Various aspects of the jet reconstruction performance have been studied using ATLAS fully detector simulated events that were produced using a variety of Monte Carlo generators. Monte Carlo generators are used to simulate events at the *truth* level, this is without including detector effects. These truth events are then sent through a GEANT [2] detector response simulation and reconstructed (including the calibration steps) with the ATLAS software, resulting in events at the so-called *fully reconstructed* or *hadronic* level. The jets obtained this way are also called *physics jets*.

Most of the studies are done by comparing fully reconstructed jets with their *matching* truth jets. In order to assign a fully reconstructed jet to a truth jet, the matching radius R_m is used to define the so-called distance in (η, ϕ) between truth and fully reconstructed jets:

$$R_m = \sqrt{\Delta\eta_{ij}^2 + \Delta\phi_{ij}^2} \tag{7.1}$$

Two jets are matched if the distance R_m in (η, ϕ) between them is smaller than a given value, e.g. $R_m \leq 0.3$.

A few variables are used to express the performance. The signal *linearity* is defined as the ratio of the energy of the reconstructed jet over its matched truth jet. The signal *uniformity* is measured by the variation of the signal as function of pseudorapidity and azimuth. Another important variable is the energy *resolution*, which is given by the width of the distribution of the relative difference between the reconstructed energy of the jet and the energy of its matched truth jet. Additional features of jet reconstruction performance are the jet reconstruction *efficiency*, defined by the ratio of the number of truth jets that are matched to a reconstructed jet over the number of truth jets, and the jet reconstruction *purity*, defined by the ratio of the number of truth jets that are matched to a reconstructed jet over the number of reconstructed jets.

It is not uncommon that for these tests, detector misalignments and other imperfections are included in the ATLAS response, so that the calorimeters do not respond optimally, but rather behave as is expected in the initial days of data taking.

7.2 Jet Finding Algorithms in ATLAS

Many jet finding algorithms exist, and four criteria are used to determine their quality [3]. A jet algorithm should be fully specified (criterion 1) and theoretically well behaved (criterion 2). Furthermore it needs to be detector independent (criterion 3) and order independent (criterion 4), the latter meaning that the algorithm has to behave equally at the parton, the hadronic and the detector levels. These last two criteria can probably never be exactly true, but must be approximately correct. But the first two criteria should be satisfied by every algorithm.

A jet algorithm is *theoretically well behaved* when it has the following features: it must be infrared safe, i.e. be insensitive to soft radiation, and collinear safe, meaning that the algorithm is independent of the fact that a certain amount of transverse momentum is carried by one particle, or by two collinear particles. Furthermore, the behavior of a theoretically well behaved jet algorithm is invariant under longitudinal boosts along the beam axis.

There are two broad groups of jet algorithms, those based in one form or another on geometrical cones in (η, ϕ) and those that involve repeated recombination of particles that are nearby in some distance measure. A recent review about jet algorithms can be found in Ref. [4].

7.2.1 Cone Algorithms

The long standing jet algorithm in ATLAS is a seeded fixed cone. This algorithm starts with ordering the input according to decreasing transverse momentum p_T. The highest p_T object gets selected if it is above a certain seed threshold (1 GeV), and all objects within a cone with fixed radius R_{cone} in (η, ϕ) around it, i.e. $\Delta R = \sqrt{\Delta\eta^2 + \Delta\phi^2} < R_{cone}$, are combined with the seed. A new direction is calculated from the four-momenta inside the initial cone and a new cone is centered around it. Objects are then (re-)collected in this new cone, and again the direction is updated. This process continues until the direction of the cone does not change anymore and the cone is labeled a jet. After that, the next seed is taken from the list and a new cone jet is formed in the same iterative way. ATLAS has chosen two cone sizes: a narrow $R_{cone} = 0.4$ and a wider $R_{cone} = 0.7$.

The jets found this way can share constituents and therefore a split and merge step after the jet formation is done. Jets which share more than a fraction f_{sm} ($f_{sm} = 0.5$ in ATLAS) of the p_T of the jets are merged, while they are split in the opposite case.

This seeded algorithm is collinear but not infrared safe; soft particles can easily modify its behavior.

A cone algorithm that does withstand the infrared problem is SISCone, a Seedless Infrared Safe Cone algorithm [5]. Contrary to the ordinary seeded cone algorithm, SISCone finds *all* stable cones using a sliding window technique, ensuring collinear and infrared safety. Cones can still be overlapping, and a split and merge step is needed. Following the recommendation of the authors, ATLAS has chosen a split-merge factor $f_{sm} = 0.75$ in order to prevent an excess of soft jets. As for the speed, SISCone takes a time $\mathcal{O}(Nn \ln n)$ to find jets among N input particles, with n the typical number of particles in a circle of radius R.

7.2.2 Sequential Recombination Algorithms

Another big class of jet finders groups the sequential recombination algorithms. For these algorithms, all pairs ij of input objects are analyzed with respect to their relative distance measures:

$$d_{ij} = \min(p_{T,i}^{2p}, p_{T,j}^{2p})\frac{\Delta R_{ij}^2}{R^2} = \min(p_{T,i}^{2p}, p_{T,j}^{2p})\frac{\Delta\eta_{ij}^2 + \Delta\phi_{ij}^2}{R^2}, \tag{7.2}$$

$$d_{iB} = p_{Ti}^{2p} \tag{7.3}$$

R is called the radius parameter [6], and p is an integer number. The index B stands for Beam. The minimum d_{min} of all d_{ij} and d_{iB} is found. If d_{min} is a d_{ij}, the corresponding objects i and j are combined into a new object k, and both elements i and j are removed from the list, while k is added to it. If d_{min} is a d_{iB}, the object i is considered to be a jet by itself and is removed from the list. This procedure is repeated until all objects are removed from the list. Contrary to the cone algorithms, no seed is needed and all objects are uniquely assigned to one jet, implying that a split and merge step is redundant and that the algorithm is infrared safe.

ATLAS uses the fast jet library [6] to run these jet finders, and two different radius parameters are implemented, namely $R = 0.4$ and 0.6. Different values of p correspond to different algorithms. For $p = 1$, one recovers the inclusive k_T algorithm [7], which will be used in Chap. 9. The case of $p = 0$ corresponds to the inclusive Cambridge/Aachen algorithm [8], and the more recent anti-k_T algorithm is obtained with $p = -1$ [9]. We will use the anti-k_T algorithm in Chaps. 10 and 11. Older implementation of these algorithms run as $\mathcal{O}(N^2)$, but the fast jet implementation runs as $\mathcal{O}(N \ln(N))$.

With anti-k_T, high p_T objects are merged first, so that the shape of the jet is conical and unaffected by soft radiation. Therefore this algorithm is very robust under pile-up conditions. The opposite is true for the inclusive k_T algorithm that starts with clustering the soft objects into the nearest hard object.

7.2.3 Anti-k_T: the ATLAS Default Algorithm

A detailed study of the performance of the jet algorithms available in ATLAS, was presented in Ref. [10]. It was found that the differences in performance are in general not very large. However, the anti-k_T algorithm is the most performant, and therefore ATLAS has decided to adopt the anti-k_T jet algorithm as default algorithm for offline reconstruction.

When considering speed and memory consumption, both the ATLAS Cone and sequential recombination algorithms perform well. SISCone, however, is a lot slower.

At low p_T the jet reconstruction efficiency and purity is algorithm dependent with anti-k_T showing the best behavior. The ATLAS Cone algorithm has the worst behavior which is due to the split and merge step. Furthermore, the ATLAS Cone is more sensitive to non-perturbative QCD effects, such as the underlying event and hadronization.

In addition, caused by the fact that the ATLAS Cone algorithm is seeded, a small but non-negligible number of events has large calorimetric energy depositions that do not belong to any jet (a detailed study of this phenomenon was done in Ref. [10]).

Both inclusive k_T and anti-k_T are stable against pile-up, due to the fact that they are infrared and collinear safe. However, since these algorithms do not use seeds, they are more sensitive to electronic noise in the calorimeter. Therefore, it is recommended to use these algorithms with input objects that are noise suppressed (topotowers or topoclusters, see Sect. 7.3).

It has also been demonstrated that the trigger efficiency can be improved by replacing the ATLAS Cone in the event filter by the anti-k_T algorithm. So far, this has not been implemented, and the trigge still uses the ATLAS Cone algorithm.

7.3 Input for Jet Finding Algorithms

The ATLAS calorimeter system has about 200,000 individual cells of various sizes. For jet finding, it is necessary to combine these cell signals into larger signal objects with massless four-momenta. Three different combinations exist: signal towers, topological cell clusters (topoclusters) and topotowers.

In the case of *signal towers*, the cells are projected onto a fixed grid in pseudorapidity and azimuth. The tower size is $\Delta\eta \times \Delta\phi = 0.1 \times 0.1$ everywhere, with about 6,400 towers in total. When the shape of a cell does not follow the shape of a given tower, only the overlapping fraction of the cell signal contributes to the tower. The towers are calibrated at the electromagnetic scale.

The major problem with towers is that all cells are included, even those cells that have a negative energy, because a too high noise subtraction has occurred during online data taking. Since a jet finder cannot deal with negative energies, negative tower signals are made positive by summing up the towers signals in the vicinity until the energy becomes positive. This method of noise cancelation has given many

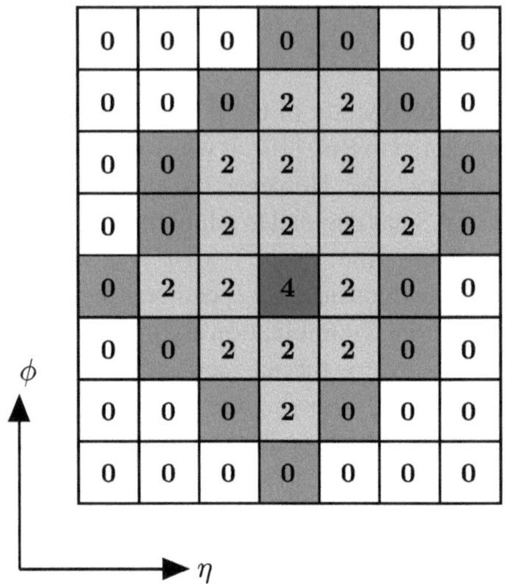

Fig. 7.1 Illustration of the topocluster formation; all *non-white* cells belong to one topocluster

problems, and a better way to deal with noise is done in the construction of topoclusters and topotowers.

Topoclusters are built according to criteria that identify energy deposits in topologically connected cells. They are an attempt to reconstruct individual parton showers. The clustering starts with seed cells with a signal-to-noise ratio Γ above a threshold S, which is currently set to four. All directly neighboring cells of these seed cells, in all three dimensions, are collected into the cluster. Neighbors of neighbors are considered for those added cells which have Γ above a secondary threshold N, with currently $N = 2$. Finally, a ring of guard cells with signal significances above a basic threshold $P = 0$ is added to the cluster. The concept is illustrated in Fig. 7.1.

After the cluster formation a split step follows, splitting clusters that have more than one local maximum. Like towers, clusters are initially formed using the basic electromagnetic scale cell signals. These clusters can already be used as input for the jet finder. In addition, clusters can be calibrated to a local hadronic energy scale first, and then used in the jet finder. Local calibration will be the topic of Sect. 7.4.2.

In the central region of the detector, clusters resolve the particle content of the jet better than towers, but because they cannot resolve individual showers in the forward regions, they cannot reproduce the jet shape very well at high pseudorapidities. In these regions, the towers perform better.

Another difference between clusters and towers, is that a tower includes all cells, while a cluster applies a noise suppression by selecting fewer cells.

The construction of *topotowers* is an attempt to combine the benefits from the geometrically well defined signal towers with the noise suppressed topoclusters. This is done by combining only those calorimeter cells that would have been selected by the topoclustering algorithm, into towers of size $\Delta\eta \times \Delta\phi = 0.1 \times 0.1$. So basically, topotowers are signal towers with noisy cells removed. Signal towers are disfavored

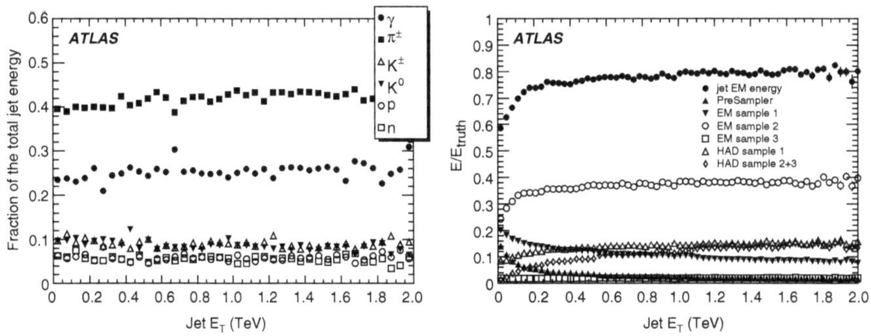

Fig. 7.2 *Left* fractional energy carried by different particle types as a function of the jet energy. *Right* fraction of true energy deposited in the different calorimeter samplings for a jet in the central ($|\eta| < 0.7$) calorimeter region as a function of its true energy. The figures are taken from Ref. [1]

with respect to topotowers. Both topotowers and topoclusters are robust against pile-up, whereas signal towers are not.

7.4 Jet Energy Calibration

7.4.1 Overview

In order to understand the underlying mechanisms of jet energy calibration, it is instructive to look at the particle content of a jet and its particle dependent energy deposits.

Figure 7.2 left shows the relative contribution of the different particle types to the jet energy as function of E_T. It can be seen that this contribution is nearly independent of the jet energy. About 40% of the total energy is carried by charged pions, while 25% is carried by photons (mainly coming from neutral pion decay), causing 25% of the energy deposits to come directly from pure electromagnetic showers.

Figure 7.2 right shows the average fractional energy deposit at the EM scale in the different calorimeter samplings with respect to the true jet energy. The full bullets represent the sum of all separate contributions. Most of the energy (about 2/3 of the reconstructed energy) is measured by the electromagnetic calorimeter. The total reconstructed energy differs significantly from the true jet energy, and this is because of a number of detector effects. The major cause is the ATLAS calorimeter non-compensation, meaning that the calorimeter response to hadrons is lower than to electrons and photons, and also non-linear with the hadron energy. Furthermore, part of the energy is lost because of dead material, cracks and gaps in the calorimeter. Finally the ATLAS magnetic field will bend low energy charged particles outside the jet cone.

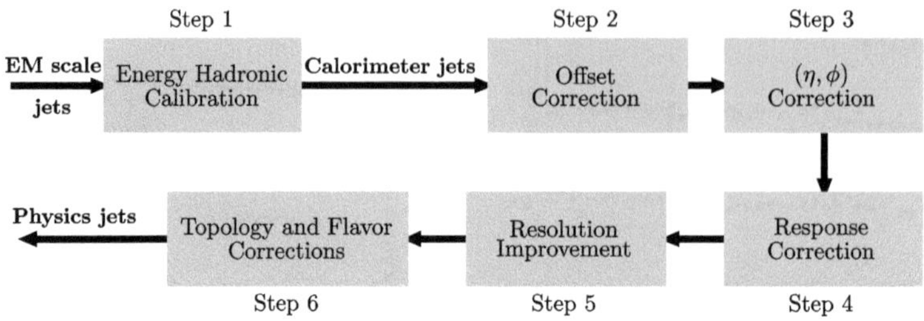

Fig. 7.3 Hadronic calibration steps

Therefore, a calibration procedure is needed to correct the jets back to the hadronic level, i.e. the scale at which hadrons are defined.

The hadronic calibration can be factorized into different steps which are shown in Fig. 7.3. First there is the actual jet energy hadronic calibration step, giving rise to so-called "Calorimeter jets". This step applies most of the corrections and its goal is to improve the jet energy resolution. However, a number of effects are not taken into account in this step: noise, pile-up, algorithm effects, etc. Therefore, step 1 is followed by additional jet-level energy corrections. An offset correction is applied in a second step: jet energy not originating from the hard scattering (electronic noise and pile-up) is subtracted. After that, a third step follows which tries to uniformize the jet response in pseudorapidity and azimuth. The fourth step consists of response corrections that finally bring the energy scale back to the hadronic (or particle) level. Next, additional corrections that improve the jet energy resolution and finally flavor corrections make step 5 and 6 respectively. We will discuss the main steps below. Important to know is that step 1 is primarily based on studies with testbeam and Monte Carlo data, while most methods mentioned in the other steps are data-driven, using suitable in-situ physics processes. After going through these steps, jets are finally calibrated at the hadronic level and referred to as "Physics jets".

The physics jets are not always the preferred analysis objects, because they are influenced by physics effects such as Initial and Final State Radiation (ISR and FSR), the underlying event, etc. Hadronic jets are therefore "cleaned" from these physics effects in a step which follows after the hadronic calibration, so that the *final* scale is reached with "Refined Physics jets" that correspond to energies at the matrix element level. It is also possible to skip step 1 and to apply the in-situ corrections directly at the EM scale.

The particles in a jet do not necessarily have high energy. An ATLAS study has shown that in order to be able to measure 95% of the energy of a 100 GeV central jet, particles of typically a few 100 MeV need to be included. Furthermore, nearly 30% of the energy of a 100 GeV jet comes from particles with energy below 10 GeV. Hence, it is important that the hadronic calibration is robust over the energy range going from about 200 MeV to a few 100 GeV.

7.4.2 Step 1: Energy Hadronic Calibration

ATLAS has four energy hadronic calibration strategies. The first one is referred to as *global cell weighting calibration*, and aims for a calibration at the jet level, after running the jet algorithm. The second strategy is called *local cell weighting* calibration and provides a calibration at the jet constituent (topocluster) level, thus before running the jet algorithm. The third method is the *longitudinal weighting method*, which makes use of the longitudinal development of the shower to correct for calorimeter-non-compensation. And the last method is the *Monte Carlo driven response correction*, which is a global (i.e. at the jet level) method using pseudorapidity and transverse momentum dependent scale factors that bring jets at the EM scale to the calibrated scale.

7.4.2.1 Global Cell Weighting

This method is based on the fact that a shower produced by a jet in the calorimeters is composed of an electromagnetic and a hadronic component. The electromagnetic component is characterized by a compact, highly dense energy deposit, while the hadronic one is broader and less dense. Going from the electromagnetic to the hadronic scale, can therefore be realized by an appropriate weighting of the calorimeter cells; cells with mainly hadronic shower activity will get a higher weight than their electromagnetic counterparts.

After running the jet algorithm, the jet energy at the electromagnetic scale is given by the summation of the calorimeter cells that make up the jet, i.e. $E_{em} = \sum_{i=cells} E_i$. In a first correction step, weights are applied so that the jet energy and momentum at the hadronic scale are given by:

$$E = \sum_{i=cells} w_i E_i \quad \text{and} \quad \vec{P} = \sum_{i=cells} w_i \vec{P_i} \tag{7.4}$$

E_i and P_i are the ith cell energy and momentum at the electromagnetic scale. The weights depend on the energy cell density (E_i/V_i, with V_i the cell volume) and the cell location ($\vec{X_i}$) and are obtained from a 45 parameter global fit by minimizing the resolution of fully Monte Carlo simulated jets to Monte Carlo truth jets. But the weights do not depend on the jet algorithm.

In a second step, residual non-linearities introduced by cracks and gaps and differences introduced by the use of different reconstruction algorithms are corrected by scale factors, which depend on pseudorapidity and transverse energy, so that the truth scale is finally recovered:

$$E_\delta = \rho_\delta E \quad \text{and} \quad \vec{P_\delta} = \rho_\delta \vec{P_i}, \tag{7.5}$$

where δ indicates the dependence on the jet finding algorithm.

The scale factors are obtained from numerical inversion, which is a technique where one starts with a four-vector, and tries to apply a correction factor based on the original (truth) four-vector that will bring you back to the original vector.

The derivation of these factors is based on the inverse of the response function $R(p_T^{\mathrm{reco}}) = p_T^{\mathrm{reco}}/p_T^{\mathrm{true}}$, where p_T^{reco} refers to the cell-weighted jet obtained from the previous step (cfr. Eq. 7.4). Given this response function, the calibrated p_T of the jet can be estimated as $p_T^{\mathrm{reco}}/R(p_T^{\mathrm{reco}})$.

The problem is that in bins of fixed p_T^{reco}, the response function is not Gaussian; due to the falling p_T spectrum, there will be more low-p_T^{true} jets, for any fixed p_T^{reco} bin, which causes a positive tail of the response function. On the other hand, the response function for jets of fixed p_T^{true} is Gaussian.

The *numerical inversion technique* [11] handles the non-Gaussian nature of the response function. The method starts with the derivation of $R(p_T^{\mathrm{true}})$ from the Gaussian response function in different p_T^{true} bins. From this an estimate $p_{T,\mathrm{est}}^{\mathrm{reco}}$ is made such that on average $p_{T,\mathrm{est}}^{\mathrm{reco}} = R(p_T^{\mathrm{true}}) \cdot p_T^{\mathrm{true}}$. The response function $R(p_{T,\mathrm{est}}^{\mathrm{reco}})$ is then calculated from $p_{T,\mathrm{est}}^{\mathrm{reco}}$, and the jet is calibrated through the inverse of this response function, by scaling its four-momentum with it.

Fully simulated Monte Carlo data samples have been used to evaluate the performance of the global cell weighting technique, and it has been observed that the different jet algorithms have similar behavior in terms of linearity and resolution. In the central region of the detector, a linearity $\langle E_{rec}/E_{truth}\rangle$ close to one (within $\pm 1\,\%$) is obtained for jets above 100 GeV, while small deviations of not more than 3% show up at lower energies.

Also the uniformity of the response over pseudorapidity is satisfactory. Apart from a small dip at $|\eta| \approx 1.5$ (in correspondence with the gap between the hadronic central barrel and the extended barrel) and at $|\eta| \approx 3.2$ (corresponding to the crack between the end-cap and the forward calorimeters), a flat behavior of $\langle E_{rec}/E_{truth}\rangle$ versus pseudorapidity is obtained.

The energy resolution σ/E improves with increasing energy, and is best in the central region. A 10% resolution is obtained for 100 GeV seeded cone (R = 0.7) jets in the region $|\eta| < 0.4$, which drops to better than 4% for 1 TeV jets.

The global calibration method operates by default on uncalibrated (at the EM scale) topoclusters, but also signal towers and topotowers can be used. When comparing the three different inputs, no difference in linearity has been observed, but topotowers and topoclusters do have a better resolution at lower energies since they are less sensitive to electronic noise.

7.4.2.2 Local Cell Weighting

This method is only applicable on topoclusters and starts by identifying them as mainly electromagnetic, mainly hadronic or unknown, depending on cluster shape and other variables, such as cluster energy. This classification is based on predictions from GEANT simulations for charged and neutral pions (they make up the major contribution to the jet energy deposit, see Fig. 7.2). Clusters classified as hadronic receive cell weights derived from detailed simulations of charged pions in order to compensate for the different response of hadrons compared to electrons. In a second step, out-of-cluster corrections are applied to correct for energy deposits that were made inside the calorimeter but fell outside the calorimeter clusters. In a final step, dead material corrections are applied to compensate for energy deposits outside the calorimeters.

The main difference with the global cell weighting is that no matching with truth jets occurs. One of the advantages of this method is that the jet finding algorithm (Sect. 7.2) runs over the input objects which have the proper scale, in contrast to the global cell weighting approach where the scale corrections are applied after the jet is reconstructed.

Important problems with this technique are the misclassification of topoclusters and the loss of particles in dead material, especially the low energy regime, since many low energetic particles do not have sufficient energy to meet the cluster reconstruction thresholds, causing the linearity never to come closer than 3% away from unity. Also the jet energy resolution is affected by this, and is typically 20% or more worse than the global cell weighting result.

Corrections for these effects are currently under study, and promising improvements are obtained using the tracker system. The ATLAS tracker covers the region $|\eta| < 2.5$ and is able to reach a 1% level of precision within the first months of data taking. Therefore, the energy over momentum ratio (E/p) of individual isolated hadrons can be used to study the local cell weighting calibration performance. The track direction is extrapolated to the calorimeter and all clusters inside a cone with radius $\Delta R = 1.0$ are included for the calculation of E of the hadron. Selection criteria are used to reduce the contamination from the underlying event.

7.4.2.3 Longitudinal Weighting Method

An alternative global calibration approach is the *longitudinal weighting method*, which makes use of the longitudinal development of the shower to correct for calorimeter non-compensation. On average, the early part of a hadronic shower is dominated by electromagnetic energy; about 70% of the energy deposited in the first interaction length of the calorimeter is electromagnetic. This fraction decreases with interaction length—it is about 25% at six interaction lengths—meaning that more of the hadron shower goes undetected. The longitudinal weighting method assigns weights to the energy deposition which are a function of calorimeter depth. The weights are pseudorapidity and energy dependent and, in order to take into account the different shower behavior of the different particles that make up the jet, they depend on the fraction of energy deposited in the liquid argon electromagnetic calorimeter as well. Although the resolution improvement is smaller with respect to other methods, the longitudinal weighting method is less demanding in terms of detector simulation dependences, compared to the previous two methods.

7.4.2.4 Monte Carlo Driven Response Correction

The Monte Carlo driven response correction [11] uses scale factors that depend on transverse momentum and pseudorapidity, to correct jets calibrated at the EM scale to the hadronic level. The scale factors are derived from numerical inversion techniques, in the same way as is done in the second step of the global cell weighting method. Of course, the scale factors one needs to apply on EM scale jets are not the same as for cell-weighted jets. The resolution obtained with this method is worse than the other methods, especially for central high-p_T jets. On the other hand, this method does not rely as much as the others on the description of the calorimeter response.

7.4.2.5 Default Method for ATLAS 2010 Data

The first three methods discussed above, heavily depend on the Monte Carlo simulation of the calorimeter response. In case the calorimeters are understood very well, these methods will give a very good performance in terms of jet energy resolution. The Monte Carlo driven response technique does not use the detailed description of the calorimeters, because the scale factors depend on η and p_T only, and no information about the calorimeter cells that make up the jet is used. This is not the case in e.g. the global cell weighting method, where the weighting coefficients largely depend on the calorimeter cell location.

Since in the early stage of ATLAS data taking, the calorimeter response might not be fully understood, it has been decided to use the Monte Carlo driven response technique as default calibration for physics studies in 2010.

7.4.3 Step 2: Offset Correction

There are two approaches to correct for jet energy not associated with the primary interaction; the first one is an average event-based correction, while the second one is a jet-by-jet correction. Jet-by-jet corrections can be used to improve the jet resolution, and they will be discussed in step 5 (see Sect. 7.4.6). Here, we will concentrate on the event-based correction.

Sources of jet energy offset are noise from the calorimeters, out-of-time pile-up (due to the fact that the liquid-argon calorimeter drift time is about 600 ns), in-time pile-up (multiple proton–proton collisions within the same bunch[1]) and the underlying event (multiple parton–parton interactions, ISR, FSR, etc). No attempt is made to correct for the underlying event as these effects are present at the jet truth level, but the other contributions—noise and pile-up—can be corrected for in an

[1] Note that the LHC at nominal luminosity is expected to have almost 25 simultaneous collisions per bunch crossing.

average way and measured as a function of pseudorapidity, instantaneous luminosity and number of interaction vertices.

Currently, ATLAS has established two alternative methods, one based on calculating the average energy in randomly placed jet cones, the other one measuring the average tower energy in zero bias and minimum bias events. The energy obtained from zero bias events can be used to estimate the noise and out-of-time pile-up, while the minimum bias events are used to estimate the in-time pile-up. The corrections obtained depend on the instantaneous luminosity, the (average) number of primary vertices and the pseudorapidity.

7.4.4 Step 3: (η, ϕ) Correction

QCD dijet events allow to check and correct the uniformity of the calibration as a function of azimuth ϕ and pseudorapidity η. The uniformity in ϕ can be checked by studying QCD jet rates which are supposed to be constant in ϕ. The uniformity in η with respect to a reference region around $\eta = 0$ can be validated using the transverse momentum balance between the dijet system. To limit unwanted imbalance effects coming from initial and final state radiation (ISR and FSR), the two jets are usually required to be back-to-back.

7.4.5 Step 4: Response Correction

7.4.5.1 Low-p_T Jets

After a uniform detector response is obtained, the absolute hadronic energy scale will be studied using $(\gamma + jet)$ or $(Z + jet)$ events, in which the p_T balance between the boson and the jet will be used to relate the hadronic scale to the well understood energy of the electromagnetic objects. The Z boson is observed via its decay into two leptons.

The $(\gamma + jet)$ balance has the advantage of higher statistics compared to $(Z + jet)$, but has the disadvantage that there is a significant QCD background from misidentified jets which mainly shows up at low p_T.

The balance is affected by various physics effects such as ISR and FSR, the underlying event, jet splitting and out-of-cone losses, which systematically limit the precision of the procedure. These effects can be as large as 5–10% at 20 GeV, and tend to decrease to the percent level at about 100 GeV. In order to reduce the bias due to additional radiation, the boson and jet are typically required to be back-to-back within $\Delta\phi$ of ± 0.2, and no further jets above a certain threshold are allowed in the event.

Alternatively, the missing $\vec{E}_T$ projection method can be used with (γ + jet) or $(Z + jet)$ events. This method projects the transverse momenta of all activity in the calorimeter and of the photon or Z onto the direction of the photon or Z, and derives a correction factor from this which is independent of the underlying event and pile-up, and to a high extent also independent of the jet algorithm.

7.4.5.2 High-p_T Jets

At very high energies, the statistics of (γ/Z+ jet) events vanishes, and in that energy region ($p_T > 400$ GeV) the hadronic scale will be evaluated using QCD multijet events by balancing the momentum of the highest jet to the momentum of the sum of the other remnants jets. The method works iteratively its way up in p_T, it is therefore also called the "bootstrap method".

Major systematic errors arise from the underlying event, from the fact that a fraction of the soft radiation is missed and from the jet energy scale uncertainty of the remnant jets. An uncertainty of the jet energy scale in the p_T-range of $400 < p_T < 1100$ GeV obtained with this method is about 8%.

Alternatively, instead of momentum balancing, the angle between particles in the jets can be used in the track angle method for high-p_T validation. The idea behind this method is that the invariant mass of two particles in a jet is approximately constant (given by the scale of Λ_{QCD}), which leads to a p_T^{-1} behavior of the $\eta - \phi$ distance between two particles in a jet (p_T the transverse momentum of the jet).

This knowledge is used in the following way. First the distance between particles in low-p_T jets (with well calibrated p_T) is determined in both data and Monte Carlo samples. In a second step, the Monte Carlo simulation is scaled in order to match the data. Then, the distance between particles in high-p_T jets in data is measured, and consequently compared with the—scaled—Monte Carlo in order to deduct the p_T of the jet.

7.4.6 Step 5: Resolution Improvement

7.4.6.1 Track-Based Corrections

For both the global and the local calibration scheme, the jet energy resolution worsens at low energy. *Track-based corrections* can be applied on jets after the standard jet energy calibrations have been implemented, improving the jet energy resolution with about 20% at low energy (about 50 GeV).

Pions are the main component of jets that deposit energy in the hadronic calorimeter. Roughly two thirds of the outgoing pions are charged, meaning that two thirds of the jet energy is carried by tracks associated with the jet. Hence, the ratio of track to calorimeter transverse momentum ($f_{track} = p_T^{tracks}/p_T^{calorimeter}$) is centered around 0.66. f_{track} provides a measure of the particle composition of the jet; jets with a large f_{track} have a larger amount of their energy carried by charged hadrons.

In bins of f_{track}, the distribution $(p_T^{reco} - p_T^{true})$ has a Gaussian shape with a mean that depends on f_{track}. This means that the combination of all f_{track}-bins will result in a much broader jet resolution compared to the individual ones. Therefore Monte Carlo data are used to study the jet response E_T^{jet}/E_T^{true} in bins of f_{track} and jet p_T, which is then used to derive correction factors. This correction to jets will reduce the f_{track}-dependence on the mean of $(p_T^{reco} - p_T^{true})$ and will therefore improve the resolution.

7.4.6.2 Longitudinal Weighting Method

The longitudinal weighting method was introduced in Sect. 7.4.2 as an alternative hadronic energy calibration method, starting from jets calibrated at the electromagnetic scale. However, longitudinal weighting can also be used to correct the response after the global or local cell weighting calibration has been applied. This correction will also improve the jet energy resolution.

7.4.7 Step 6: Topology and Flavor Corrections

Another improvement of the jet calibration can be obtained specifically for jets originating from a b quark. When a b quark is created, it decays semileptonically into a muon and a neutrino with roughly a 10% probability. The neutrinos carry away a fraction of the energy, introducing a systematic underestimation of the jet energy. But since these jets can be tagged by the presence of a muon, the jet energy scale can be corrected for this as a function of jet and muon p_T, leading to a noticeable improvement of the jet response of b-jets.

References

1. ATLAS Collaboration, Expected Performance of the ATLAS Experiment: Detector, Trigger and Physics. CERN, Geneva (2008)
2. Geant4 Collaboration, Geant4-a simulation toolkit. nuclear instruments and methods in physics research section A: accelerators, pectrometers, detectors and associated equipment. **506**(3), 250–303 (2003)
3. Blazey G.C. et al. Run II Jet Physics in *Proceedings of the Run II QCD and Weak Boson Physics Workshop*, (2000), (hep-ex/0005012)
4. G.P. Salam, Towards jetography (2009) (hep-ph/0906.1833)
5. G.P. Salam, G. Soyez, A practical seedless infrared-safe cone jet algorithm. JHEP **086**, (2007)
6. M. Cacciari, G.P. Salam, Dispelling the N3 myth for the Kt jet-finder. Phys. Lett. **B641**, 57 (2006) (hep-ph/0512210)
7. S. Catani, Y.L. Dokshitzer, M.H. Seymour, B.R. Webber, Longitudinallyinvariant k⊥-clustering algorithms for hadron–hadron collisions. Nucl. Phys. B, **406**, 187–224 (1993)

8. L. Dokshitzer Y, G.D. Leder, S. Moretti, B.R. Webber, Better jet clustering algorithms. JHEP **1997**, 001 (1997)
9. M. Cacciari, G.P. Salam, G. Soyez, The anti-k_t jet clustering algorithm. JHEP **2008**(04), 063 (2008)
10. L. Asquith et al., Technical Report ATL-COM-PHYS-2009-630 CERN, Geneva (2009)
11. D. Lopez Mateos, E.W. Hughes, A. Schwartzman. A Simple pT- and η-Dependent Monte Carlo-Based Jet Calibration. Technical Report ATL-PHYS-INT-2009-077 CERN, Geneva (2009)

Chapter 8
Jet Reconstruction with 2010 ATLAS Data

8.1 Jet Algorithm and Jet Calibration

From the previous chapter it is clear that many jet algorithms and calibration schemes are foreseen in ATLAS. Due to the absence of real data for many years, the jet reconstruction performance has mostly been evaluated using Monte Carlo generated events. Even though these studies are very important for our understanding, real data studies form undoubtedly a vital ingredient of the performance evaluation.

Now that the LHC has become operational, all methods will be investigated and improved. But for the first *physics* publications in 2010, it has been decided to only work with jets that are constructed from topoclusters using the anti-k_T algorithm, making the jets theoretically well behaved and robust against calorimeter noise.

Several calibration steps were discussed in the previous chapter, going from the jet energy calibration, over offset and $(\eta - \phi)$ corrections, to the response correction and corrections to improve the resolution. But with early data, only the first step, i.e. the jet energy calibration will be applied to the jets, and none of the other steps, since these steps require more data. The very first jets will be corrected using the Monte Carlo driven response technique (see Sect. 7.4.2), since this method does not depend heavily on the Monte Carlo description of the ATLAS calorimeter system.

8.2 Data Quality Requirements and Event Cleaning

8.2.1 Run Selection

ATLAS collects its data in so-called data runs, of which many typically cover a few hours. These runs are subdivided into luminosity blocks, corresponding to approximately two minutes of data taking with a constant instantaneous luminosity.

N. Boelaert, *Dijet Angular Distributions in Proton–Proton Collisions*, Springer Theses, 99
DOI: 10.1007/978-3-642-24597-8_8, © Springer-Verlag Berlin Heidelberg 2012

Using the data quality framework that was discussed in Sect. 6.4, green, yellow[1] and red data quality flags are assigned to all sub-systems of the detector and all luminosity blocks.

The ATLAS collaboration has decided to have the same data quality requirements for jet studies as for studies with missing transverse energy (E_T^{miss}). The data sample for generic jet and E_T^{miss} studies is obtained by requiring a selection of data quality flags to be green, and applying a few other criteria [1].

Since the primary sub-systems for jet studies are the calorimeters (LAr and TileCal), these systems are required to be green. Note that the presence of hot and noisy cells does not prohibit green data quality flags, because these fake cells can be eliminated using jet cleaning cuts. However, a large LHC beam background will mark the LAr as bad, since this results in large background levels seen by the LAr.

Furthermore, the trigger system must be functional, and the magnet system must be on and stable. Also the muon system must be fully operational, since the missing E_T reconstruction depends on it. Because many jet studies use the properties of tracks related to jets, the inner detector is required to operate at nominal voltage and without major problems. Data quality flags for basic jet and E_T^{miss} distributions aimed at finding problems in the jet reconstruction, are required to be green as well. A final requirement is the presence of *stable* beams as declared by the LHC.

8.2.2 Event Selection

ATLAS defines a *collision event* as a head-on inelastic collision, which is the type of event suitable for many physics studies. The main backgrounds to collision events are caused by collisions between one proton beam and the beam gas within the beam pipe (*beam-gas collisions*) and muons or pions traveling in the halo of the beam (*beam-halo*). There is also a small background component formed by cosmic muons overlapping with the actual collision events.

The background events are removed from the collision events by requiring that the event has passed an appropriate trigger, which for jet studies is either a jet trigger or a minimum bias trigger. The minimum bias trigger consists of Minimum Bias Trigger Scintillators (MBTSs): two discs of 16 scintillators situated on each side of the ATLAS detector, outside the end-caps of the inner detector. The minimum bias trigger requires at least one hit on each MBTS, together with a signal from beam pick-up timing devices (BPTXs), which are located on either side of the ATLAS detector, at a distance of 175 m along the beam pipe. Jet triggers look for single or multi jets with a minimum jet energy.

Furthermore, the timing of energy deposits in the end-cap inner wheels and forward calorimeters is considered as well when trying to remove backgrounds; on *both* sides there needs to be a number of cells that have deposits well above the noise level.

---

1 Right now, the yellow flag is assigned to bad but recoverable systems, i.e. systems that are expected to obtain green flags after a—final—reprocessing of the data (with improved reconstruction software). Therefore yellow flags should not appear in the final dataset.

On top of these requirements, only events with at least one vertex with at least five reconstructed tracks and a longitudinal position from the center of ATLAS < 10 cm are selected.

8.2.3 Jet Selection

Fake jets arise from various sources, ranging from hardware problems, LHC beam conditions, cosmic rays and cosmic air showers. Fake jets are classified as bad and ugly jets.

Bad jets arise from in-time fake energy depositions, and they need to be rejected at the analysis level. Dedicated cleaning cuts have been developed for this purpose:

- In order to reduce noise in the hadronic end-caps (HECs), either the fraction of the jet's energy deposited in the HEC is required to be smaller than 0.8, or the number of energy-ordered cells accounting for at least 90% of the jet energy is supposed to be greater than five.
- Coherent noise affecting the EM-calorimeters is a rare but unfortunate event that fakes bad jets. These bad jets are characterized by a poor quality of the calorimeter readout signal in the cells or by a large amount of energy reconstructed by the EM cells. Therefore jets are required to have a limited number of EM cells with bad quality or to have a fraction of energy in the EM-calorimeter that is smaller than 0.95.
- Jets reconstructed from out-of-time energy depositions, for example due to photons produced by cosmic ray muons, need elimination as well. This can be done by cutting on the jet time, which is defined as the energy-weighted cell time average within two beam crossings. Jets with a jet time $>$ 50 ns are disregarded. This is a rather loose cut, given that the bunch crossing time is 25 ns (but so far, filled bunches are much more separated).

Ugly jets are real jets but with a wrong calibration. This happens for jets that have a large fraction of their energy from masked cells, i.e. cells that are broken, so that only an estimate of the energy deposit can be made based on the neighboring cells. When the fraction of energy from problematic cells to the total jet energy (at the EM scale) exceeds 0.5, the jet is labeled ugly. Also jets which have an energy fraction in the gap between the hadronic barrel and end-cap that exceeds 0.5, are labeled ugly because the scintillators in that region have not been fully commissioned.

8.3 Jet Reconstruction Performance

8.3.1 Jet Energy Scale Uncertainty

The first determination of the jet energy scale (JES) and the evaluation of its systematic uncertainty for inclusive jets have been described in Ref. [2]. The systematic uncertainty on the jet energy scale is assumed to be given by a Gaussian distribution with a width $\sigma(p_T, \eta)$ depending on the transverse momentum and the pseudorapidity of the jet.

Monte Carlo studies show that the Monte Carlo driven response technique restores the jet energy and transverse momentum response within 2% for non-isolated jets. Furthermore, using this jet energy calibration method, a good agreement between Monte Carlo simulations and data has been found, assuring the applicability of this method to data.

Deviations from unity for the jet response are caused by the fact that for the derivation of the calibration constants, a different jet selection and topology is used and the assumption is made that each constituent needs the same average compensation. Therefore, the non-linearity of the jet response needs to be added to the jet energy scale uncertainty.

Other contributions to the uncertainty can be divided into different categories. Firstly there are the detector related contributions, such as the uncertainty in the detector description (dead material, etc), shifted beamspots, noise description, the hadronic shower model and the uncertainty on the EM-scale (see Sect. 7.4.1 for the definition of the EM-scale).

Secondly there are the theoretical uncertainties which stem from uncertainties in the fragmentation and the underlying event models, and the Monte Carlo event generator tune.

Finally also multiple proton-proton collisions (pile-up) contribute to an uncertainty, since—so far—there is no explicit correction for the extra energy they cause during the jet calibration.

Dedicated Monte Carlo samples with systematic variations that aim at providing a conservative estimate of the detector-related and theoretical uncertainties have been generated and compared to the nominal sample (used to derive the calibration constants) in order to deduct a systematic uncertainty for each individual contribution. The uncertainties from non-independent effects are grouped (e.g. the uncertainty on the hadronic shower description and the noise thresholds) and the largest deviation is taken bin-by-bin. These results and other non-factorizable uncertainties are then combined in quadrature. The non-linearity of the jet response is added linearly.

An extra uncertainty is added in order to account for the uncertainty on the relative response of jets in the hadronic end-cap compared to jets in the barrel. This choice is motivated by a better knowledge of the material in the barrel than in the end-cap. The relative response in data of end-cap to central jets is evaluated using the p_T balance technique (see Sect. 7.4.4) and the results show that the deviation from unity of the relative energy scale in the end-cap region is nearly negligible for $|\eta| < 2.1$ and does not exceed 2% for $2.1 < |\eta| < 2.8$).

Furthermore the difference between data and Monte Carlo in the relative energy scale has been evaluated to be 2.4% for jets with $2.1 < |\eta| < 2.8$.

Both contributions are added as an uncertainty for jets with $|\eta| > 1.2$. This choice over-estimates the JES uncertainty because it partially double-counts the contribution from the dead material uncertainty, but it is considered as a conservative estimate in

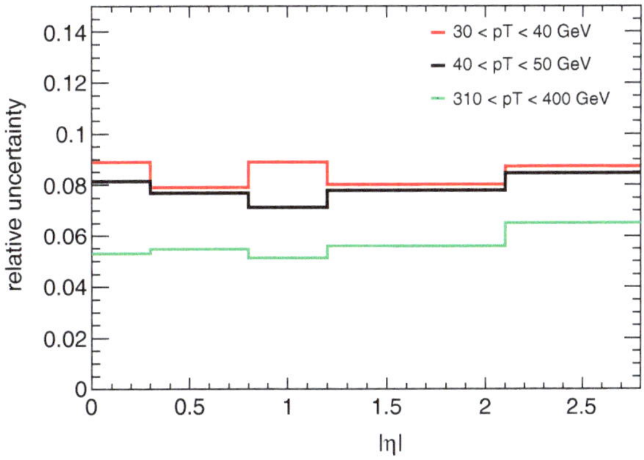

Fig. 8.1 Absolute JES uncertainty $\sigma(p_T, \eta)$ as function of η for three different bins of jet p_T.

a region where the material budget could not be well described by the Monte Carlo simulation.

As for the pile-up contributions, again a conservative approach is taken; the average impact of pile-up is considered as an additional systematic uncertainty estimated directly from data, assuming that all jets and events are affected. This provides an upper bound on the uncertainty due to pile-up effects. For data that were recorded up to May 2010, and for jets with $p_T = 20$ GeV, the pile-up relative systematic uncertainty is about 1% in the barrel and $1-2\%$ in the end-cap. For $p_T > 50$ GeV, the pile-up uncertainty is only significant for $|\eta| > 2.1$ and it is smaller than 1%.

The total inclusive uncertainty, obtained from combining all contributions, is shown in Fig. 8.1 for three different p_T bins as a function of pseudorapidity.

References

1. M. Baak, M. Petteni, N. Makovec, Data-quality requirements and event cleaning for jets and missing transverse energy reconstruction with the ATLAS detector in proton-proton collisions at a center-of-mass energy of $\sqrt{s} = 7$ TeV. Technical report ATL-COM-PHYS-2010-247, CERN, Geneva (2010)
2. ATLAS Collaboration. Jet energy scale and its systematic uncertainty in ATLAS for jets produced in proton-proton collisions at $\sqrt{s} = 7$ TeV. Technical report ATL-COM-PHYS-2010-404, CERN, Geneva (2010)

Chapter 9
Dijet Angular Distributions at $\sqrt{s} = 14$ TeV: A Phenomenology study

9.1 Introduction

The LHC is designed to collide protons at a center of mass energy of 14 TeV. This energy will however only be reached after a long initial run at $\sqrt{s} = 7$ TeV. Even though collisions at nominal energy are still far ahead in the future, it is very instructive to already explore at this moment the ultimate potential of the LHC at nominal energy.

In this chapter we will therefore present a phenomenology study at $\sqrt{s} = 14$ TeV. First we will perform a QCD study; we will calculate the distributions up to NLO in four different bins of dijet invariant mass using different Monte Carlo programs and different jet algorithms, and we will also investigate the systematic uncertainties coming from the choice of the parton distribution functions and the renormalization and factorization scales.

In the second part we will present the effects on the distributions coming from gravitational scattering and black hole formation in large extra dimensions. Assuming a 25% systematic uncertainty, we will report a discovery region for the mass bin $1 < M_{jj} < 2$ TeV at 10 pb^{-1} integrated luminosity.

This study was published in Ref. [1] and reported in Ref. [2].

9.2 Kinematics Cuts

Dijet angular distributions were discussed in Sect. 4.2. The goal is to study the differential cross section

$$\frac{\mathrm{d}\sigma}{\mathrm{d}\chi} = \int \mathrm{d}x_1 \int \mathrm{d}x_2 f_1(x_1, Q^2) f_2(x_2, Q^2) \frac{\mathrm{d}\hat{\sigma}}{\mathrm{d}\chi} \tag{9.1}$$

in bins of dijet invariant mass. QCD is expected to give rise to a cross section that is rather flat in χ, while new physics often shows up as an increase of the cross section

DOI: 10.1007/978-3-642-24597-8_9,

Table 9.1 Values of $p_{T,\min}$ for 4 different dijet invariant mass bins and 2 values of $\chi_{\max}$

Mass bin	$\chi_{\max} = 100$ pT$_{\min}$ (GeV)	$\chi_{\max} = 600$ pT$_{\min}$ (GeV)
$0.5 < M_{jj} < 1$ TeV	35	14
$1 < M_{jj} < 2$ TeV	70	28
$2 < M_{jj} < 3$ TeV	140	57
3 TeV $< M_{jj}$	210	86

at low χ. The following four mass bins were chosen for $\sqrt{s} = 14$ TeV: $0.5 < M_{jj} < 1$ TeV, $1 < M_{jj} < 2$ TeV, $2 < M_{jj} < 3$ TeV and 3 TeV $< M_{jj}$.

Selection cuts in pseudorapidity were previously discussed as well; see Eq. 4.22:

$$|\eta_1 - \eta_2| < 2\eta_{\max} - c \tag{9.2}$$

$$|\eta_1 + \eta_2| < c \tag{9.3}$$

In this study we take $c = 1.5$. The ATLAS calorimeters can measure jets fully up to $\eta_{\max} \sim 4$. With this choice for the value of c and with $\eta_{\max} = 4$, the angular distributions can be measured up to $\chi_{\max} \sim 600$. In case we are only interested in measuring up to $\chi_{\max} \sim 100$ (e.g. for new physics searches), we can limit ourselves to $\eta_{\max} = 3.1$.

At lowest order, the relation between M_{jj}, χ and the transverse momentum p_T is the following:

$$M_{jj} = p_T(\sqrt{\chi} + 1/\sqrt{\chi}) \tag{9.4}$$

The selection cuts on dijet mass ($M_{jj,\min} < M_{jj} < M_{jj,\max}$) and χ ($\chi < \chi_{\max}$), will determine the minimum p_T the two leading jets in the event need to have in order to pass the selection cuts:

$$p_{T,\min,\mathrm{LO}} = \frac{M_{jj,\min}}{(\sqrt{\chi_{\max}} + 1/\sqrt{\chi_{\max}})} \tag{9.5}$$

NLO contributions will lower the minimum transverse momentum with a factor $\sqrt{2}$, as a consequence of the fact that the subleading jet in the event can never have a transverse momentum less than half the transverse momentum of the leading jet. So, up to NLO:

$$p_{T,\min} = \frac{M_{jj,\min}}{\sqrt{2}(\sqrt{\chi_{\max}} + 1/\sqrt{\chi_{\max}})} \tag{9.6}$$

Equation 9.6 will be used as a p_T-selection cut to optimize the efficiency of the Monte Carlo event generation. Table 9.1 lists the values of $p_{T,\min}$ for the different mass bins and for two different values of $\chi_{\max}$.

9.3 QCD Calculations

We will use both JETRAD [3] and NLOJET++ [4] for NLO jet calculations. The programs use a conceptually very different approach, see Chap. 3 for a detailed discussion.

While the considerations in the previous section are at the parton level, experiments have to deal with jets made of hadrons, which are the result of three major steps: hard interaction, parton showering and hadronization. Jets carry the memory of the hard interaction and a good jet-finding algorithm can exhibit this information in an infrared and collinear safe way. A detailed discussion about jet algorithms was given in Sect. 7.2. Here we will use the inclusive k_T algorithm and SISCone. Although seeded cone algorithms are not infrared stable, we will also present results with a seeded cone algorithm because they are still frequently used in experiments. More precisely, we will use a seeded, iterative cone with progressive removal that comes with JETRAD and was brought into use by the CERN UA1 collaboration [5]. NLOJET++ comes with an exact seedless cone algorithm which was first proposed in Ref. [6] and finds all stable cones in a given configuration, ensuring infrared safety. We will also show results using this algorithm.

Note that the final state for an NLO order calculation at the matrix element level contains at most three partons and that we consider only those events with at least two jets. Therefore most problems emerging from infrared and collinear instabilities are absent in our calculations.

The major difference between the previously mentioned cone algorithms is that the JETRAD seeded cone clusters two nearby partons if their separation in (η, ϕ) is less than the cone radius R, while the NLOJET++ seedless cone and SISCone do so if the cone containing both partons is stable, i.e. if their separation is smaller than $R(1 + z)$, with $z = p_{T,2}/p_{T,1}$ and $p_{T,2} < p_{T,1}$ [7].

Figure 9.1 shows the angular distributions for the mass bin $1 < M_{jj} < 2$ TeV, calculated both with JETRAD and NLOJET++, using an inclusive k_T algorithm with radius parameter R = 1.0. JETRAD uses a different parametrization of the strong coupling constant than NLOJET++, which explains the difference between the curves to a large extent. To illustrate this, we show in Fig. 9.2 the angular distributions with α_s kept constant at 0.1.

NLOJET++ (coded in C++) has the advantage over JETRAD (coded in fortran) that it can be combined with more modern jet algorithms, such as SISCone and an exact seedless cone. But it has the disadvantage that the angular distributions have a statistical error that is not homogeneous over the whole χ range, but is instead increasing with χ, which is not the case for a JETRAD calculation, as can be seen from Fig. 9.1.

Figure 9.3 compares calculations at the Born (lowest order) level with next-to-leading order calculations, done with JETRAD, for the four different mass bins and for $1 < \chi < 600$. All NLO distributions have been calculated with the next-to-leading order CTEQ6M PDF [8], while the leading order distributions have been calculated both with the CTEQ6M PDF and with the leading order CTEQ6L1

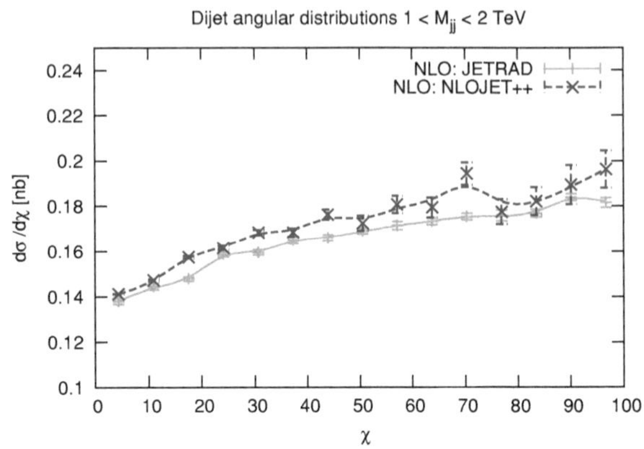

Fig. 9.1 NLO calculations using JETRAD and NLOJET++ and an inclusive k_T algorithm with R = 1.0 for the mass bin $1 < M_{jj} < 2$ TeV

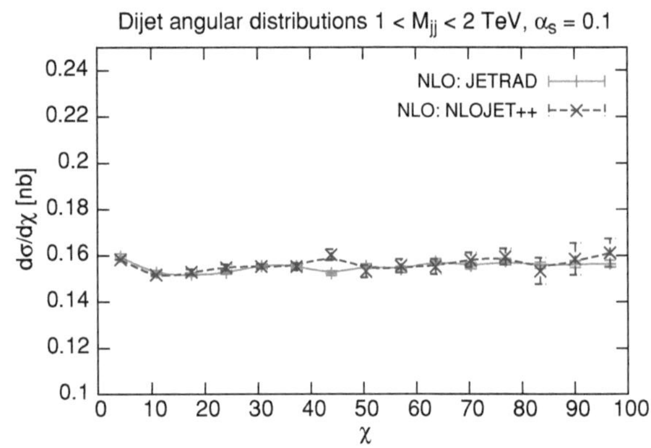

Fig. 9.2 NLO calculations using JETRAD and NLOJET++ and an inclusive k_T algorithm with R = 1.0 for the mass bin $1 < M_{jj} < 2$ TeV and for $\alpha_s = 01$

PDF. Two different jet algorithms are used for the NLO calculations; the JETRAD seeded cone algorithm with radius R = 0.7, and an inclusive k_T algorithm with R = 1.0 level calculation, the outgoing partons are back-to-back, so that a jet algorithm is redundant. The NLO angular distributions with the two different jet algorithms tend to have the same shape, but differ in absolute normalization.

From this figure, we also note that the NLO calculations are flatter than the Born calculations, especially at high χ values, which can be explained to a large extent by the fact that the running of α_s with p_T (or equivalently χ) has more effect on a LO than an NLO calculation. To illustrate this, we have plotted in Fig. 9.4 four different LO calculations in the mass bin $1 < M_{jj} < 2$ TeV; the dashed blue curve is the same Born calculation as the one presented in the top right plot of Fig. 9.3. For this calculation, the values of α_s and of the factorization scale (μ_F) have been varied according to the p_T of the hardest jet. The other three curves show what happens if α_s and/or μ_F have been kept constant. The grey dotted curve is a calculation with $\alpha_s = 0.1$ and $\mu_F = 100$ GeV and is much flatter compared to the blue one. The red curve has been calculated with $\alpha_s = 0.1$ and running of μ_F, and the green one with

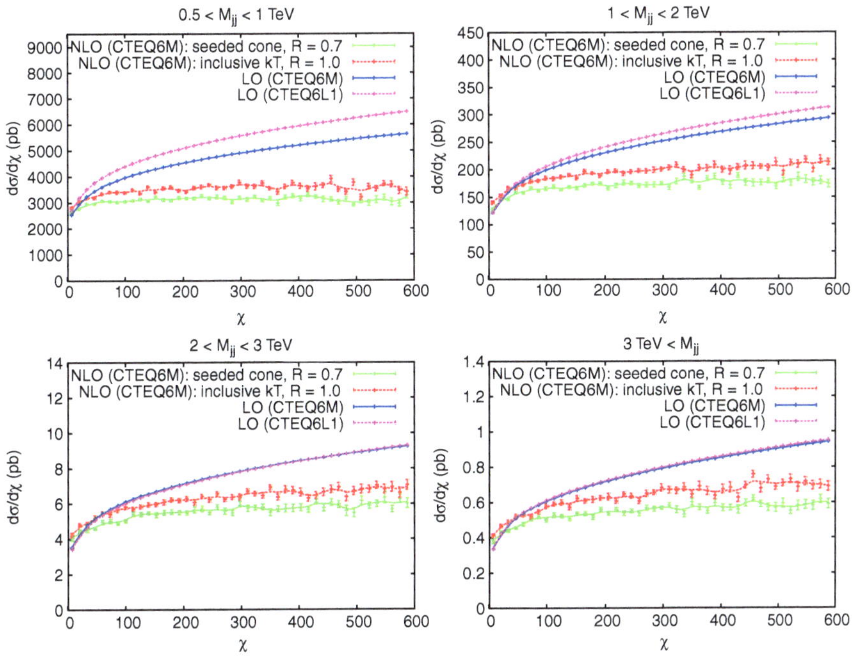

Fig. 9.3 LO and NLO angular distributions calculated with JETRAD for 4 different mass bins. The NLO calculations are reconstructed with two different jet algorithms: the JETRAD seeded cone algorithm with R = 0.7, and an inclusive k_T algorithm with R = 1.0. All NLO curves are calculated with the CTEQ6M PDF, while the LO distributions are calculated with both the CTEQ6M and the CTEQ6L1 PDF. In the bottom two plots, the LO CTEQ6L1 and LO CTEQ6M curves are overlapping

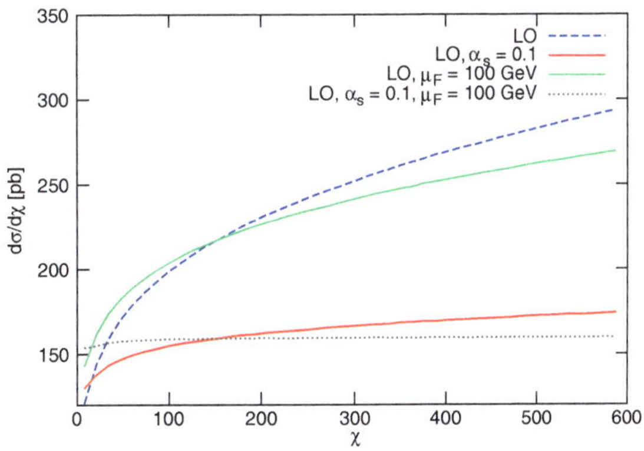

Fig. 9.4 Influence of the running of α_s and the variation of μ_F on LO distributions

$\mu_F = 100\,\text{GeV}$ and running of α_s. There is a much bigger effect on the distributions from keeping α_s fixed than from μ_F.

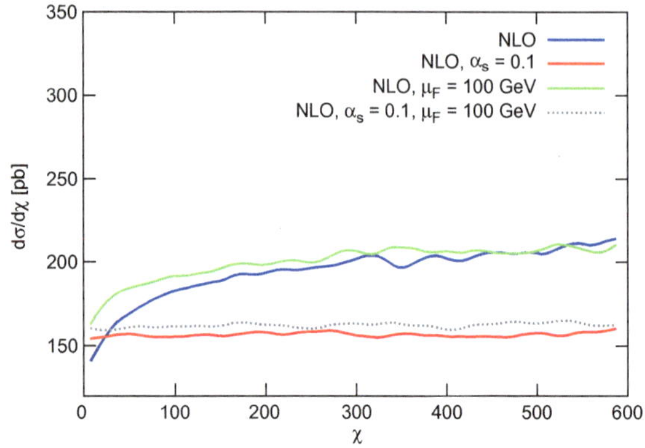

Fig. 9.5 Influence of the running of α_s and the variation of μ_F on NLO distributions. The calculations are done with JETRAD and an inclusive k_T algorithm with R = 1.0

But these are observations at the Born level. At NLO, the sensitivity due to scale variations is reduced as can be observed in Fig. 9.5. This is because perturbation theory tells us that an all-order calculation should not depend on the renormalization scale at all, and therefore, compared to the LO calculation, an NLO calculation is more stable against the running of α_s. Furthermore, an NLO order calculation uses an NLO expansion of α_s, and the running of α_s at NLO is less pronounced than at LO. Figures 9.4 and 9.5 clearly illustrate the need for an NLO order calculation with the running of α_s and μ_F enabled.

This is also confirmed by calculations done with NLOJET++. In Fig. 9.6 we present the angular distributions, done with NLOJET++, for the same mass bins, using the NLOJET++ seedless cone algorithm with R = 0.7 and overlap 0.5, the SIS-Cone algorithm with R = 0.7 and overlap 0.75, and an inclusive k_T algorithm with R = 0.6. We show the distributions for $1 < \chi < 100$ only and, compared to the big variations between the LO and NLO calculations observed at high χ ($\chi > 100$) in Fig. 9.3 , the difference between LO and NLO is much less at small values of χ.

In Fig. 9.7 we show the ratio of the NLO cross section calculated with CTEQ6M over the LO one calculated with CTEQ6L1, for $1 < \chi < 100$ and for different jet algorithms, a quantity which is often called the k-factor in literature. The k-factor varies around 1 with a deviation of 30% at the most.

The calculations above have been done with the CTEQ6M and CTEQ6L1 parton distribution functions and with a normalization and factorization scale chosen to be the transverse momentum of the hardest jet. Uncertainties coming from parton distribution functions (PDFs) and the choice of renormalization (μ_R) and factorization (μ_F) scale will contribute to a systematic error. We will investigate them using JETRAD with an inclusive k_T algorithm with R = 1.0. The exact scale and PDF uncertainties may vary with different jet algorithms and cone sizes.

Figure 9.8 shows the angular distributions for the mass bin $1 < M_{jj} < 2$ TeV for three different PDF-sets, namely CTEQ6M, CTEQ66 [9] and MSTW2008NLO [10]. The distributions differ mainly in absolute normalization, and less in shape.

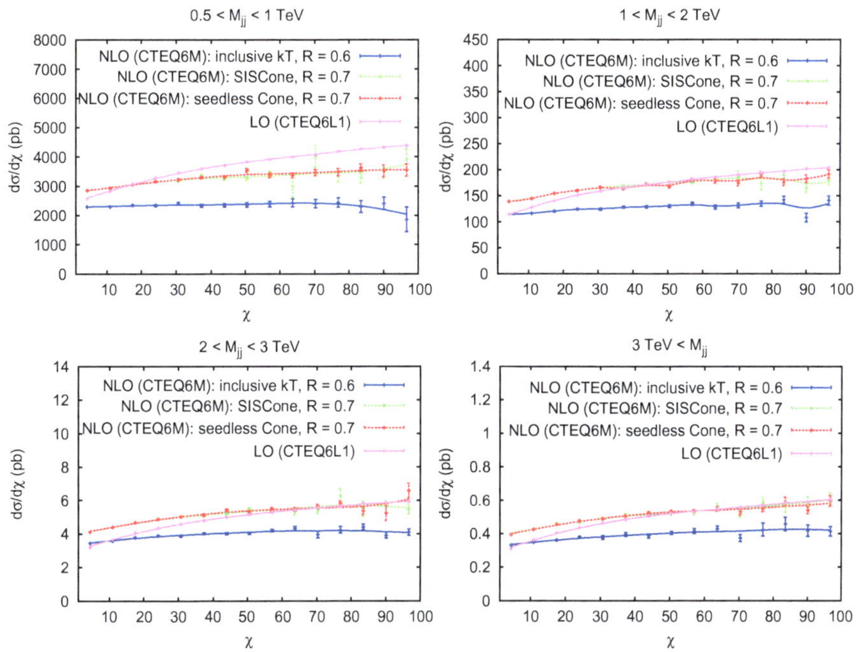

Fig. 9.6 LO and NLO angular distributions calculated with NLOJET++ for 4 different mass bins. The NLO calculations are reconstructed with three different jet algorithms: an inclusive k_T algorithm with R = 0.6, SISCone with R = 0.7, and the NLOJET++ seedless cone algorithm with R = 0.7

Normalizing the distributions to unit area make them differ no more than 3% over the whole χ range, as can be seen in Fig. 9.9.

To further examine the uncertainties coming from PDFs, we have calculated the angular distributions for all 44 error members of the CTEQ66 PDF and applied the Master Equation suggested in Ref. [11] to deduct a positive (+) and negative (−) uncertainty on a quantity X:

$$\Delta X^{+}_{\max} = \sqrt{\sum_{i=1}^{N}[\max(X_i^{+} - X_0, X_i^{-} - X_0, 0)]^2} \tag{9.7}$$

$$\Delta X^{-}_{\max} = \sqrt{\sum_{i=1}^{N}[\max(X_0 - X_i^{+}, X_0 - X_i^{-}, 0)]^2} \tag{9.8}$$

ΔX^{+} adds in quadrature the PDF error contributions that lead to an increase in the observable X, and ΔX^{-} the PDF error contributions that lead to a decrease. Using this formula, the error on the dijet angular distribution calculated with the central PDF member, in the mass bin $1 < M_{jj} < 2$ TeV is plotted in Fig. 9.10 as a blue band

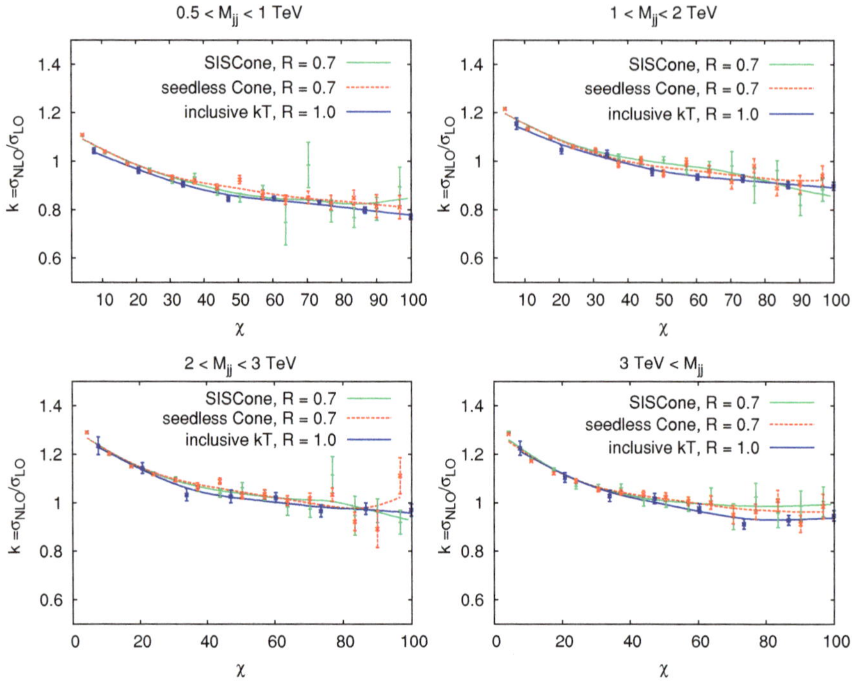

Fig. 9.7 Ratio of σ_{NLO} (CTEQ6M) over σ_{LO} (CTEQ6L1) for 4 different mass bins calculated with NLOJET++

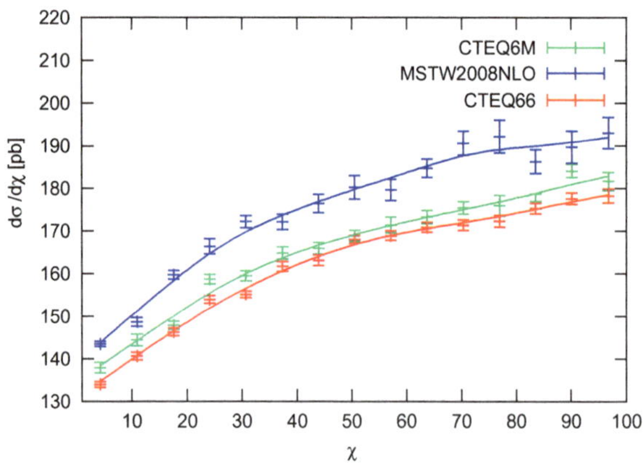

Fig. 9.8 Angular distributions with different PDF-sets for the mass bin $1 < M_{jj} < 2$ TeV. The calculations are done with JETRAD and an inclusive k_T algorithm with R = 1.0

around the central member. The uncertainties given in terms of percentage are shown in Fig. 9.11.

The choice of μ_R and μ_F will also influence the distributions. We have studied this by letting μ_R and μ_F vary independently between 0.5, 1 and 2 times the transverse momentum of the hardest jet, resulting in 9 different distributions in total.

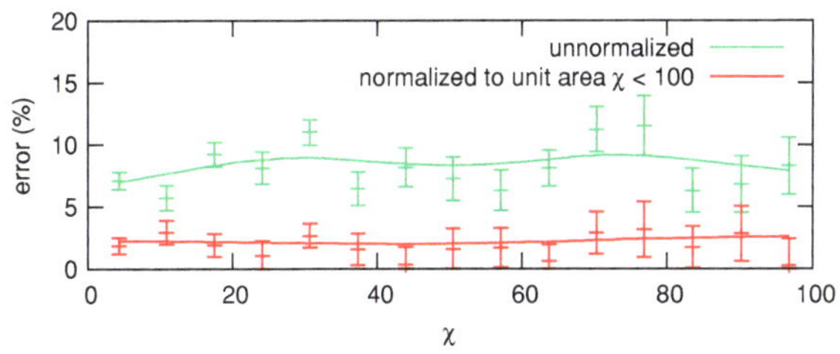

Fig. 9.9 Difference in terms of percentage between the dijet angular distributions calculated with different PDFs (see Fig. 9.8) in the mass bin $1 < M_{jj} < 2$ TeV. The differences have been calculated both for the distributions normalized to unit area and not normalized

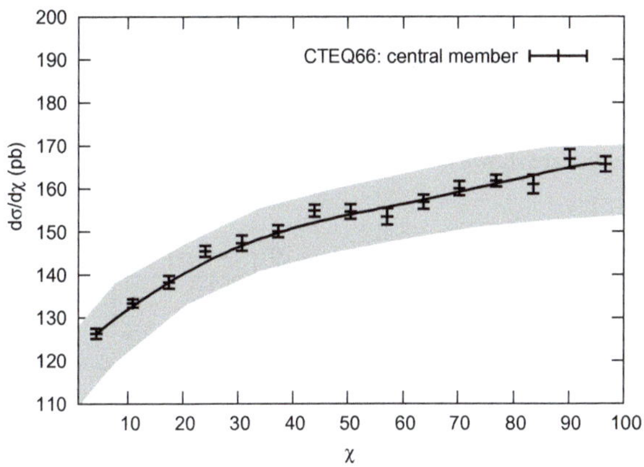

Fig. 9.10 Central member of the CTEQ66 PDF, together with its uncertainty band for the mass bin $1 < M_{jj} < 2$ TeV. The calculations are done with JETRAD and an inclusive k_T algorithm with R = 1.0

Figures 9.12 and 9.13 summarize the results for the mass bin $1 < M_{jj} < 2$ TeV. The other mass bins have similar results. For all these figures, the plots at the left show the distributions for $\chi < 100$, while the plots at the right go up to $\chi = 600$. The variables r and f used in the figures, denote the fraction of the transverse momentum of the leading jet at which respectively the renormalization and factorization scale are evaluated. In Fig. 9.12 μ_F is kept constant, while μ_R is varied, which influences both the shape and the normalization of the distributions. The effect is small at low χ but increases drastically with increasing χ values. The plots in Fig. 9.13 all have fixed μ_R and varying μ_F, which causes a change rather in normalization and not so much in shape.

The difference between these distributions is an estimate of the uncertainty coming from the choice of μ_R and μ_F. In Fig. 9.14 we show the uncertainty on the distributions both with and without normalization to unit area $1 < \chi < 100$. Normalizing the distributions reduces the error drastically; averaged over the whole χ

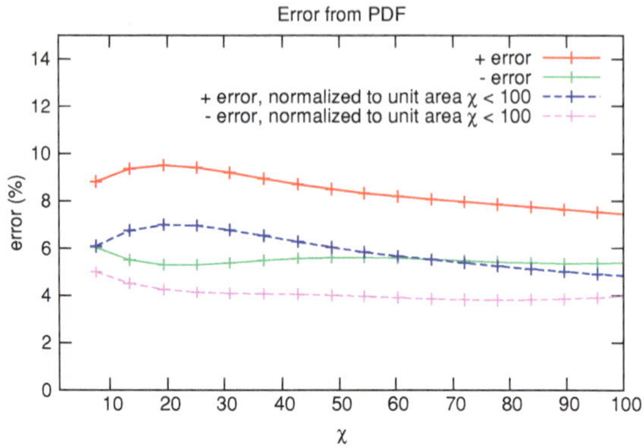

Fig. 9.11 Uncertainty in the angular distribution calculated with the central member of the CTEQ66 PDF, in the mass bin $1 < M_{jj} < 2$ TeV coming from the intrinsic uncertainty of the PDF. The + error leads to an increase of the cross section, while the - error leads to a decrease (see Eq. 9.8) in the text. The uncertainties are shown on the curves both with and without normalization to unit area $1 < \chi < 100$

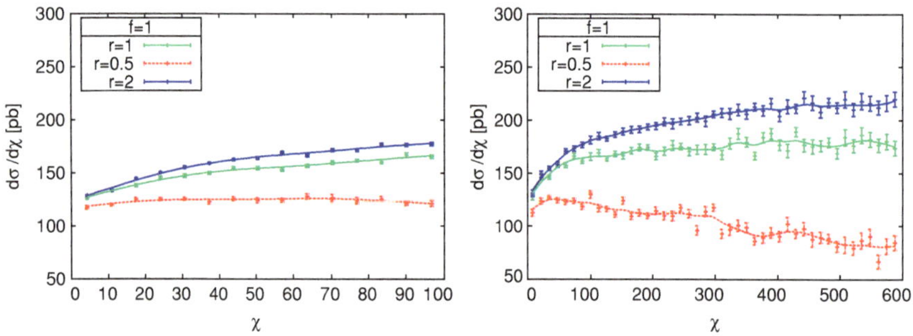

Fig. 9.12 Variation of the choice of μ_R for the mass bin $1 < M_{jj} < 2$ TeV. The variables r and f in the figure denote the fraction of the transverse momentum of the leading jet at which respectively the renormalization and factorization scale are evaluated, i.e. $\mu_R = rp_T$ and $\mu_F = fp_T$. *Left* $\chi < 100$, *right* $\chi < 600$. The calculations are done with JETRAD and an inclusive k_T algorithm with R = 1.0

range, the uncertainty on the normalized distribution is 9%, and is never exceeding 20%.

Figure 9.15 shows the combination in quadrature of the uncertainties coming from the choice of renormalization and factorization scale (Fig. 9.14), together with the intrinsic uncertainty from the CTEQ66 PDF (Fig. 9.11), drawn as an error band around the calculation done with nominal values (central CTEQ66 member and $\mu_R = \mu_F = p_T$ of the leading jet). Both the distributions with (plot right) and without (plot left) normalization to unit area $1 < \chi < 100$ are shown. In both plots, the error band is dominated by the error coming from the choice of the scales, more precisely, the renormalization scale introduces the major uncertainty.

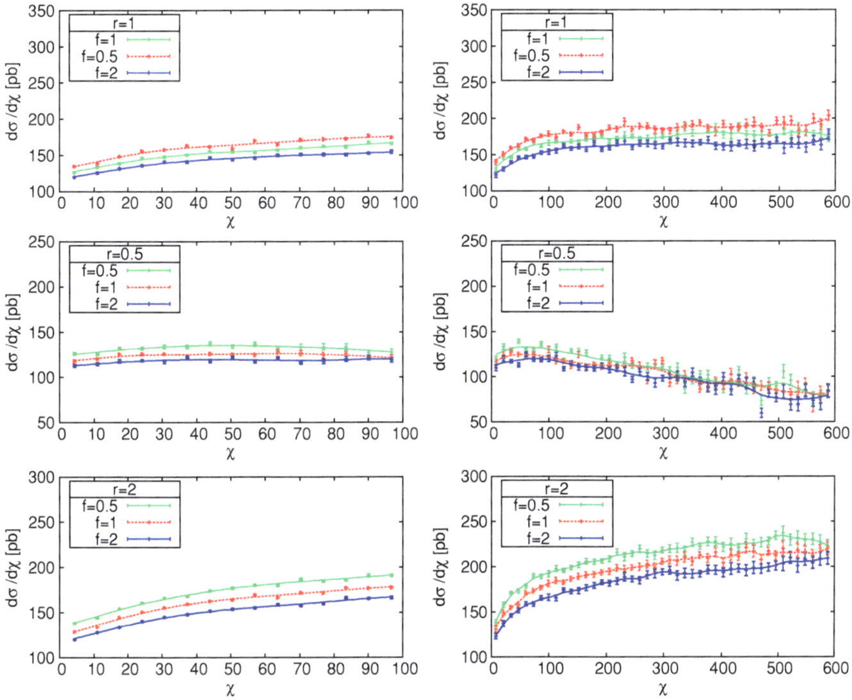

Fig. 9.13 Variation of the choice of μ_F for the mass bin $1 < M_{jj} < 2$ TeV. The variables *r* and *f* in the figure denote the fraction of the transverse momentum of the leading jet at which respectively the renormalization and factorization scale are evaluated, i.e. $\mu_R = rp_T$ and $\mu_F = fp_T$. *Left* $\chi < 100$, *right* $\chi < 600$. The calculations are done with JETRAD and an inclusive k_T algorithm with R = 1.0

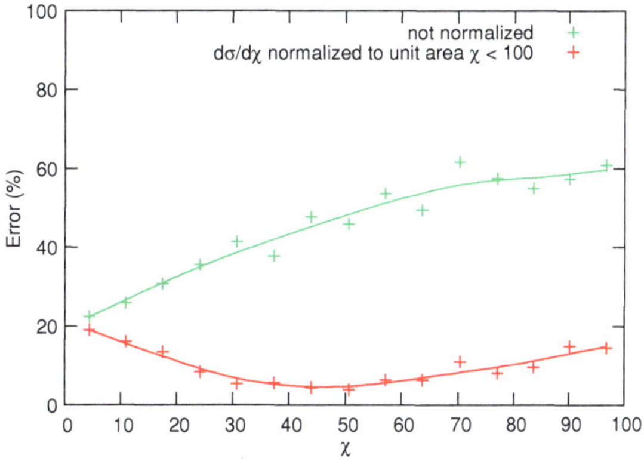

Fig. 9.14 Systematic uncertainty coming from the choice of $\mu_{R,F}$ for the mass bin $1 < M_{jj} < 2$ TeV, both on normalized and not normalized distributions

When normalized to unit area $1 < \chi < 100$, the combined error does not exceed 20% over the whole χ range.

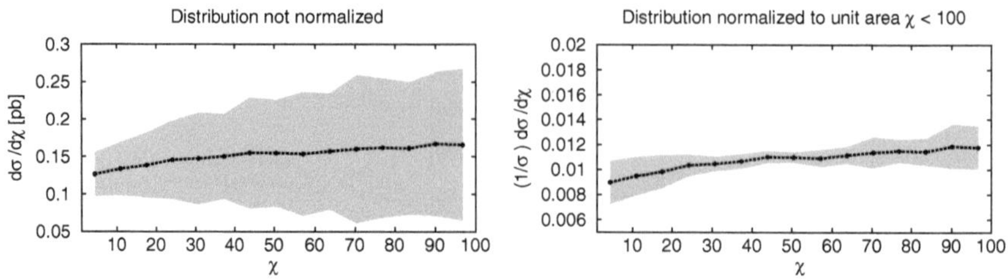

Fig. 9.15 *Black line*: calculation done with nominal central CTEQ66 member and $\mu_R = \mu_F = p_T$ of the leading jet, both with (*plot right*) and without (*plot left*) normalization to unit area $1 < \chi < 100$. *Blue band*: error band from combining the uncertainties coming from the choice of renormalization and factorization scale (Fig. 9.14) , together with the intrinsic uncertainty from the CTEQ66 PDF (Fig. 9.11)

The calculations have been done at the parton level of the hard interaction, without showering, multiple interactions or hadronization. In Fig. 9.16 we show what happens if we turn on these processes. In the top figure, we have plotted angular distributions calculated with PYTHIA 6.410 [12]. The dashed gray curve is a calculation at the Born level, without switching on showering, multiple interactions or hadronization. Initial state radiation (ISR) and primordial k_T (prim kT) have been turned on for the dashed red curve, and the full blue curve is a calculation with final state radiation (FSR) included (but initial state radiation and primordial k_T turned off). As can be observed in the figure, final state radiation causes losses out of the jet cone. The dashed magenta curve covers both initial and final state showers, primordial k_T and multiple interactions. Finally, the full green curve includes all these processes and hadronization. The calculations are done with the PYCELL cluster jet algorithm from the PYTHIA library and the PYTHIA default settings for initial and final state radiation, multiple interactions and hadronization. For this particular choice of jet algorithm and settings, the difference between the calculation at the Born level and the one at the hadron level is non negligible, as can be seen from their ratio, shown in the bottom plot of Fig. 9.16. The corrections may vary for different jet algorithms and PYTHIA settings.

9.4 Gravitational Scattering and Black Hole Formation in Large Extra Dimensions

In this section we use the GravADD generator to make a study about gravitational scattering and black hole formation in large extra dimensions. We refer to Chap. 5 for more information about the model and the generator.

The calculations have been done using an inclusive k_T algorithm with separation parameter 1.0. We have used the k-factor which was derived in the previous section (see Fig. 9.7) to scale the PYTHIA QCD distributions up to NLO.

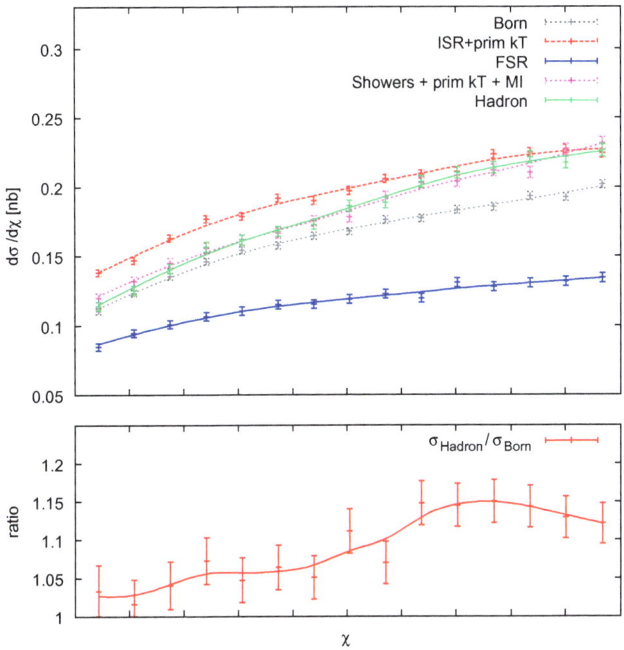

Fig. 9.16 Including primordial k_T, initial and final state radiation, multiple interactions and hadronization in a LO calculation. The calculations are done using PYTHIA and the PYCELL cluster jet algorithm

Table 9.2 Different parameter sets

Name	M_{eff} (TeV)	n	M_s/M_P	M_P (TeV)	M_s (TeV)
C1	1.0	6	1.0	0.282	0.282
C2	1.0	6	2.0	0.564	1.128
C3	1.0	6	4.0	1.128	4.513
C4	0.5	6	8.0	1.128	9.027
C5	1.0	6	8.0	2.257	18.05
C6	1.0	4	4.0	1.263	5.053

As discussed in Chap. 5 , the model parameters determining the phenomenology are M_P, n and M_s. An equivalent set of parameters is M_{eff} (Eq. 5.35), M_s/M_P and n, and we will work with the latter one. Table 9.2 shows a few combinations. All parameter choices in Table 9.2 have been simulated in proton-antiproton collisions at an energy of $\sqrt{s} = 1.8$ TeV for the mass bin 635 GeV $< M_{jj}$ and for $\chi \leq 20$, and turned out to show no deviations from QCD, which is consistent with the dijet angular distribution measurements done by the CDF [13] and D0 collaboration [14] at the Tevatron.

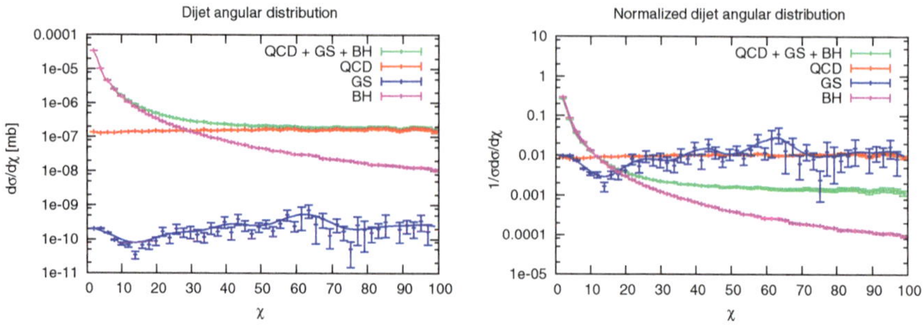

Fig. 9.17 Angular distributions for C1 (see Table. 9.2) for the mass bin $1 < M_{jj} < 2$ TeV. *Left* cross section in mb, *right* cross section normalized to unit area $1 < \chi < 100$

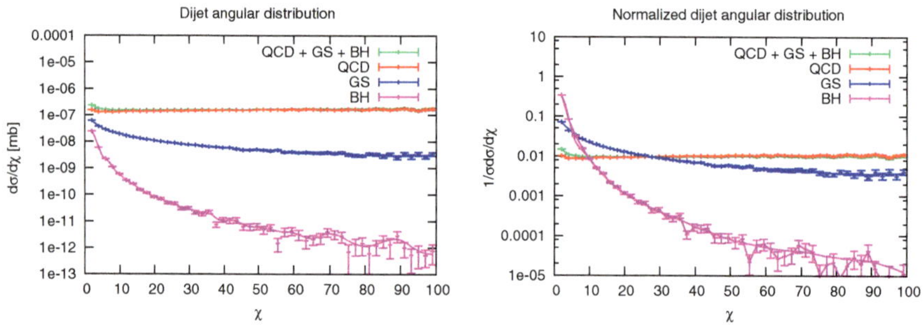

Fig. 9.18 Angular distributions for C3 (see Table. 9.2) for the mass bin $1 < M_{jj} < 2$ TeV. *Left* cross section in mb, *right* cross section normalized to unit area $1 < \chi < 100$

To get a feeling of the impact of the parameters, we have performed a few runs with PYTHIA with parton showers, multiple interactions and hadronization turned off. Figures 9.17, 9.18, 9.19, 9.20, 9.21 and 9.22 show the dijet angular distributions, with and without normalizing to unit area $1 < \chi < 100$, for some of the parameter sets defined in Table. 9.2. Previous experiments [13, 14] have shown that normalizing the distributions reduces the experimental error (see also Sect. 10.6 of the next chapter). Furthermore, in the previous section we have demonstrated that the theoretical error on normalized distributions is smaller as well (Fig. 9.15). We have plotted the different contributions—i.e. gravitational scattering (GS), black holes (BH) and QCD—separately, as well as their sum. As with most new physics, the biggest effects from gravitational scattering and black hole formation show up at low χ (high p_T) values. How much each process contributes to the total cross section, depends on the parameter settings and the mass bin. We will make a few quantitative observations.

Increasing M_s/M_P while keeping M_{eff} fixed, results in an increase of gravitational scattering and a decrease of black hole formation. The effect is clearly visible when set C1 (Fig. 9.17) is compared with set C3 (Fig. 9.18) for the mass bin $1 < M_{jj} < 2$ TeV. For C1, the total cross section at low χ values, is dominated by black holes, while

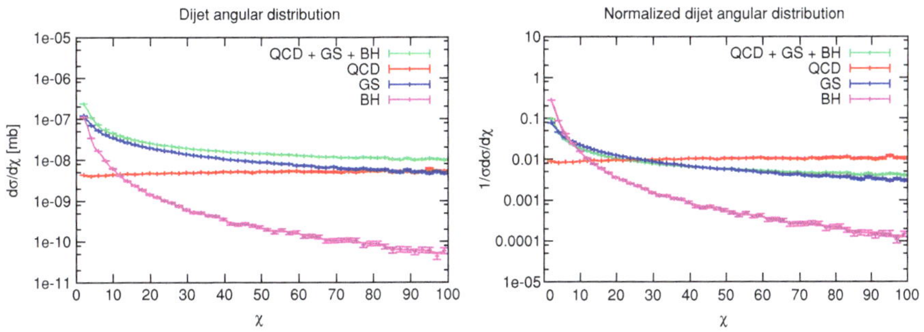

Fig. 9.19 Angular distributions for C3 (see Table. 9.2) for the mass bin $2 < M_{jj} < 3$ TeV. *Left* cross section in mb, *right* cross section normalized to unit area $1 < \chi < 100$

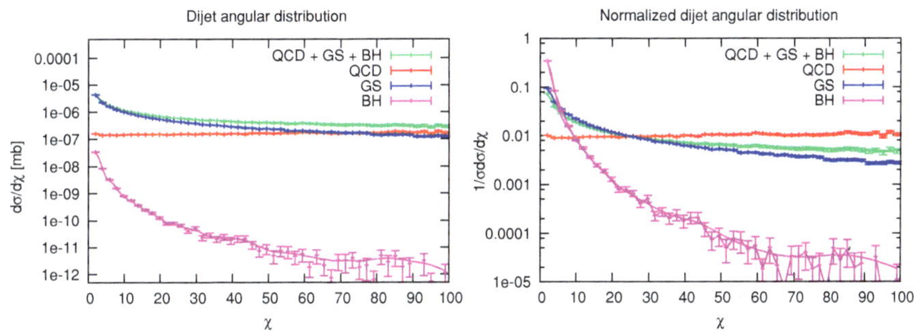

Fig. 9.20 Angular distributions for C4 (see Table. 9.2) for the mass bin $1 < M_{jj} < 2$ TeV. *Left* cross section in mb, *right* cross section normalized to unit area $1 < \chi < 100$

gravitational scattering is of no importance. But this changes drastically for C3; black holes have almost completely disappeared for C3, but the gravitational scattering cross section has increased by two orders of magnitude. QCD still dominates in this mass bin, but gravitational scattering starts to dominate the cross section for mass bins $\geq$2 TeV, as can be seen in Fig. 9.19.

Compared to C3, C4 has the same fundamental Planck scale and number of extra dimensions (1 and 6 TeV, respectively), but a different width of the membrane (9 vs 4.5 TeV), and this causes effects from gravitational scattering to set in at lower mass values; from Fig. 9.20, it is observed that gravitational scattering dominates the cross section from 1 TeV onwards, which is not the case for C3 (Fig. 9.18).

Figure 9.21 examines what happens if we double the effective and fundamental Planck scale; we have plotted the angular distribution in the mass bin 3 TeV $< M_{jj}$ for C4 (Fig. 9.21 left) and C5 (Fig. 9.21 right). As expected, gravitational scattering becomes weaker for C5, but it still strongly dominates QCD. The cross section for black hole formation has decreased much more than the one for gravitational scattering.

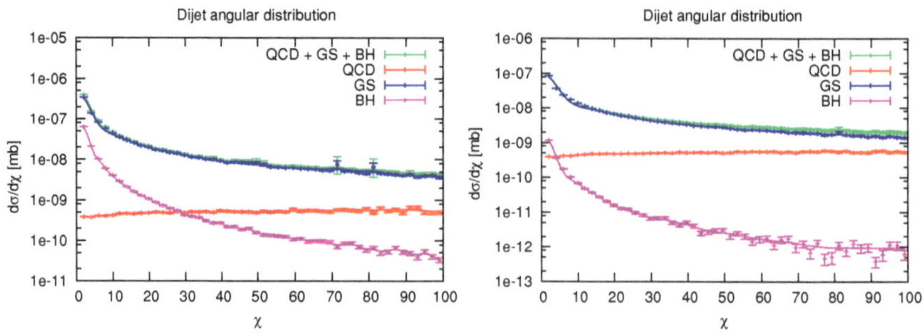

Fig. 9.21 Angular distributions (in mb) in the mass bin 3 TeV $< M_{jj}$ for C4 (*left*) and C5 (*right*)

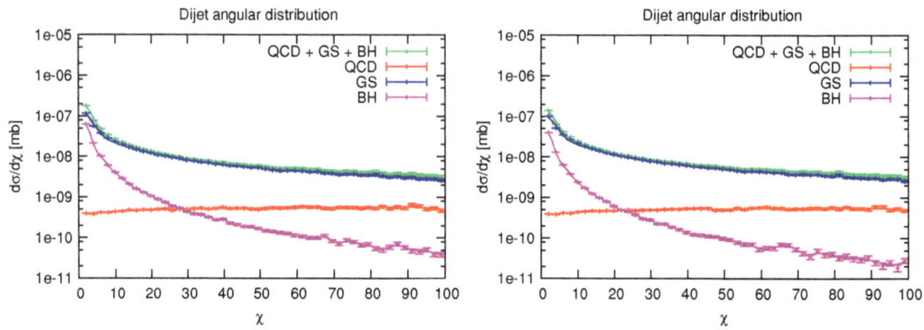

Fig. 9.22 Angular distributions (in mb) in the mass bin 3 TeV $< M_{jj}$ for C3 (*left*) and C6 (*right*)

Finally also the number of dimensions matters. Figure 9.22 gives the angular distributions in the mass bin 3 TeV $< M_{jj}$ for 6 (parameter set C3, plot at the left) and 4 (parameter set C6, plot at the right) extra dimensions, while keeping the other parameters constant. The difference between 6 and 4 extra dimensions is very small.

Note that when gravitational scattering is the most important contribution, deviations from the QCD cross section are still visible at large χ values, which is not the case for black holes. But because at large χ values gravitational scattering is mainly a t-channel process, the shape of the graviational scattering curve is close to the QCD one, and the difference between QCD and gravitational scattering disappears in the normalized distributions.

Deviations from QCD are most visible at low χ values, and a summary of the above observations is done in Table. 9.3 here we give for a selection of parameter settings and mass bins, the integrated cross sections between $\chi = 1$ and $\chi = 50$ for QCD (σ_{QCD}), gravitational scattering (σ_{GS}) and black holes (σ_{BH}) separately.

We will now compare the different new physics scenarios with QCD for a given integrated luminosity, so that we can establish a region sensitive to discovery. We use the distributions that are normalized to $\chi < 50$ and perform a chi-square (χ^2) test between them.

Table 9.3 Different parameter sets and the relevance of GS and BH in several mass bins

Name	M_{eff}(TeV)	n	M_s/M_P	M_P (TeV)	M_s (TeV)	Mass bin (TeV)	σ_{QCD} (nb)	σ_{GS} (nb) (nb)	σ_{BH} (nb)
C1	1.0	6	1.0	0.282	0.282	[1,2]	7.23	8.2310^{-3}	113
C2	1.0	6	2.0	0.564	1.128	[1,2]	7.23	4.7410^{-2}	20.5
C3	1.0	6	4.0	1.128	4.513	[1,2]	7.23	7.0110^{-1}	1.0110^{-1}
C3	1.0	6	4.0	1.128	4.513	[3,14]	2.3010^{-2}	8.8010^{-1}	2.3310^{-1}
C4	0.5	6	8.0	1.128	9.027	[0.5,1]	47.0	3.44	2.7810^{-3}
C4	0.5	6	8.0	1.128	9.027	[1,2]	7.23	37.37	0.10
C4	0.5	6	8.0	1.128	9.027	[2,3]	6.9610^{-2}	2.98	1.2210^{-1}
C4	0.5	6	8.0	1.128	9.027	[3,14]	2.3010^{-2}	1.89	2.3310^{-1}
C5	1.0	6	8.0	2.257	18.05	[3,14]	2.3010^{-2}	5.4310^{-1}	4.1510^{-3}
C6	1.0	4	4.0	1.263	5.053	[3,14]	2.3010^{-2}	8.1510^{-1}	1.4510^{-1}

We have used the following recipe. For a given integrated luminosity, we consider the normalized distributions $(dN_{\text{QCD}}/d\chi)/N_{\text{QCD}}$ and $(dN_{\text{total}}/d\chi)/N_{\text{total}}$, with N_{QCD} and N_{total} respectively the number of QCD and total (= QCD + GS + BH) events. We perform a chi-square (χ^2) test between these distributions to test the null hypothesis that $(dN_{\text{total}}/d\chi)/N_{\text{total}}$ follows the QCD distribution. We use both a statistical and systematic uncertainty for the calculation of χ^2:

$$\chi^2 = \sum_{\text{allbins}\,i} \left[\frac{\left(\frac{N_{\text{QCD},i}}{N_{\text{QCD}}} - \frac{N_{\text{total},i}}{N_{\text{total}}} \right)^2}{s^2_{\text{stat},i} + s^2_{\text{sys},i}} \right], \tag{9.9}$$

where $N_{\text{QCD},i}$ and $N_{\text{total},i}$ are the number of QCD and total events respectively in bin i. The statistical error $s_{\text{stat},i}$ is for each bin taken as $\sqrt{N_{\text{QCD},i}}$. See Appendix A for the derivation of the above formula (Eq. A.21).

Based on the fact that the theoretical uncertainty does not exceed 20% (see Fig. 9.15 right), and that experimental uncertainties reported by the Tevatron experiments are less than 11% [13, 14], the systematic error $s_{\text{sys},i}$ is taken to be 25% over the whole χ range. Using χ^2 (Eq. 9.9) and the number of degrees of freedom, the probability p of having $dN_{\text{total}}/d\chi/N_{\text{total}}$ given the null hypothesis is true, can be calculated. The null hypothesis of identity is rejected for $p < 0.1$.

Let us focus on the mass bin $1 < M_{jj} < 2$ TeV and consider only those worlds with n = 6 extra dimensions and work with an integrated luminosity of 10 pb^{-1}. We have used the recipe mentioned above to test several physics scenarios and the green colored area in Fig. 9.23 shows for which model parameters—M_{eff} and M_s/M_P—the null hypothesis is rejected. This region is from now on called the region of discovery.

As can be seen from the figure, large values of M_{eff} ($M_{\text{eff}} > 1.5$ TeV) and small values of M_s/M_P ($M_s/M_P < 1$) fall outside the region of discovery. The reason is the absence of black hole creation because in that region the lower limit on the black

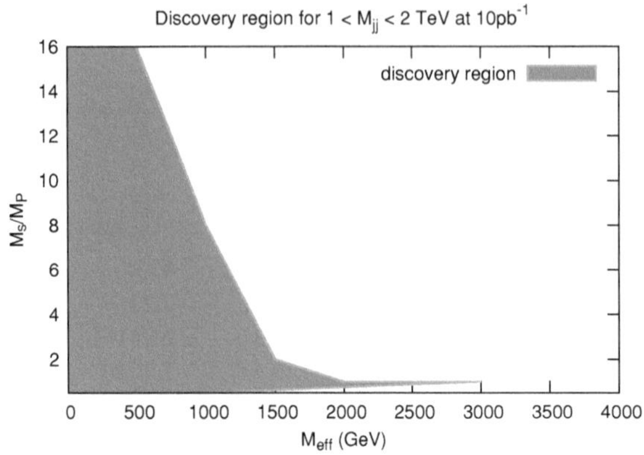

Fig. 9.23 Discovery region for the mass bin $1 < M_{jj} < 2\,\text{TeV}$ at $10\,\text{pb}^{-1}$, assuming a 20% systematic uncertainty

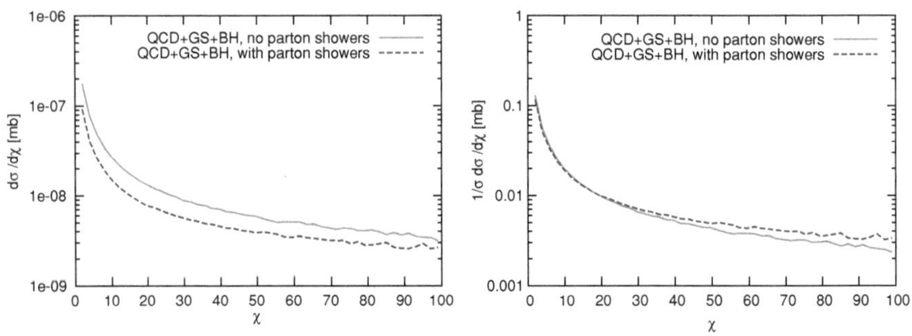

Fig. 9.24 Angular distributions in the mass bin $3\,\text{TeV} < M_{jj}$ for C3, with and without parton showers. *Left* cross section in mb, *right* cross section normalized to unit area $1 < \chi < 100$

hole mass is drastically increasing with decreasing M_s/M_P. See Eq. 5.40 and the discussion underneath.

Above calculations have been done without parton showering, multiple interactions or hadronization. In Fig. 9.24 we compare the angular distributions with and without parton showers in the mass $3\,\text{TeV} < M_{jj}$ bin for C3. The effect of parton showers is most visible at low χ values.

9.5 Conclusions

We have discussed dijet angular distributions at $\sqrt{s} = 14\,\text{TeV}$. First we have performed a QCD study, then we have shown the distributions in a scenario with gravitational scattering and black hole formation in large extra dimensions.

We have used JETRAD and NLOJET++ for the calculation of QCD up to next-to-leading order, and found that both programs generate similar results. The angular distributions at NLO are flatter than the Born calculations, especially at large values of χ ($\chi > 100$), which is mainly caused by the fact that the running of α_s has less effect on NLO than on LO calculations. Different jet algorithms tend to keep the shape of the distributions unchanged, but give a different normalization.

We have investigated the systematic uncertainties coming from the choice of renormalization (μ_R) and factorization (μ_F) scale and parton distribution function (PDF), and found that a change in μ_F and PDF mainly influences the normalization. On the other hand, a change in μ_R has an impact on both the normalization and the shape of the distributions; the distributions have a similar behavior at low χ but tend to spread out as χ increases.

The effects on dijet angular distributions from gravitational scattering and black hole production, have been studied in the ADD scenario, with the extra requirement that the membrane on which the standard model fields are allowed to propagate, has a finite but small width. The model parameters are the fundamental Planck scale, the width of the membrane and the number of extra dimensions, and it has turned out that the phenomenology is very much dependent on the fundamental Planck scale and the width of the membrane. For a fundamental Planck scale of around 1 TeV and for a wide range of parameter settings, quantum gravity effects are noticeable in mass bins above 1 TeV. For small widths of the membrane, gravitational scattering is the most important process, while black hole formation dominates for wider membranes. In both cases, the effects mainly show up at small values of χ. The same conclusions can be made for the normalized distributions. Using the shape of the distributions, rather than the absolute normalization, for $\chi < 50$, we have determined the region of parameter space that could be discovered with 10 pb^{-1} integrated luminosity and a 25% systematic uncertainty in the mass bin $1 < M_{jj} < 2$ TeV.

In conclusion, uncertainties from QCD that cannot be reduced by normalizing the distributions mainly show up at large values of χ ($\chi > 100$), while effects from quantum gravity are mostly present at small values of χ. Depending on the region of χ under study, dijet angular distributions can therefore either be used as a probe for new physics or as a test of QCD.

References

1. N. Boelaert, T. Åkesson, Dijet angular distributions at $\sqrt{s} = 14$ TeV.. EPJ C **66**, 343–357 (2010) (hep-ph/09053961)
2. N. Boelaert, Dijet angular distributions at $\sqrt{s} = 14$ TeV. EPS-HEP. in *Proceedings of Science*, 2009
3. W.T. Giele, E.W.N. Glover, D.A. Kosower, Higher order corrections to jet cross sections in hadron colliders. Nucl. Phys. **B403**, 633 (1993) (hep-ph/9302225)
4. Z. Nagy, Next-to-leading order calculation of three-jet observables in hadron–hadron collision. Phys. Rev. **D68**, 094002 (2003) (hep-ph/0307268)

5. G. Arnison et al., Hadronic jet production at the CERN proton - anti-proton collider. Phys. Lett. B **132**, 214 (1983) UA1 Collaboration,
6. Blazey G.C. et al. Run II Jet Physics in *Proceedings of the Run II QCD and Weak Boson Physics Workshop*, 2000, (hep-ex/0005012)
7. C.M. Harris, P. Richardson, B.R. Webber, CHARYBDIS: A black hole event generator. JHEP **0308**, 033 (2003) (hep-ph/0307305)
8. J. Pumplin, D.R. Stump, J. Huston, H.L. Lai, P. Nadolsky, W.K. Tung, New generation of parton distributions with uncertainties from global QCD analysis. JHEP **0207**, 012 (2002) (hep-ph/0201195)
9. P.M. Nadolsky, H.L. Lai, Q.-H. Cao, J. Huston, J. Pumplin, D. Stump, W.K. Tung, P.C. Yuan, Implications of CTEQ global analysis for collider observables. Phys. Rev. **D78**, 013004 (2008)
10. A.D. Martin, W.J. Stirling, R.S. Thorne, G. Watt, Parton distributions for the LHC. EPJ **C63**, 189–285 (2009)
11. T. LeCompte, Level 1 Jet Efficiency Measurements Using the Tag and Probe Method Technical Report ATL-COM-PHYS-2007-0143. (CERN, Geneva, 2007)
12. T. Sjöstrand, S. Mrenna, P. Skands, PYTHIA 6.4 physics and manual. JHEP **0605**, 026 (2006) (hep-ph/0603175)
13. F. Abe et al., Measurement of Dijet Angular Distributions by the Collider Detector at Fermilab. Phys. Rev. Lett. **77**(27), 5336–5341 (1996) CDF Collaboration
14. B. Abbott et al., Measurement of Dijet angular distributions and search for quark compositeness. Phys. Rev. Lett. **80**(4), 666–671 (1998) D0 collaboration

Chapter 10
Preparing ATLAS for the Measurement of Dijet Angular Distributions at $\sqrt{s} = 7\,\text{TeV}$

10.1 Introduction

This chapter describes a Monte Carlo study aimed at preparing ATLAS for the measurement of dijet angular distributions, by establishing the measurement procedure, investigating experimental and theoretical uncertainties, discussing unfolding issues, and studying the sensitivity to gravity mediated processes that were discussed in Chap. 5. This work was documented in Ref. [1].

Since in the initial phase of LHC protons collide at $\sqrt{s} = 7\,\text{TeV}$, we carry out this study at this energy and with an integrated luminosity ranging from $1\,\text{pb}^{-1}$ to $10\,\text{pb}^{-1}$, which is a realistic expectation for the first year of data taking.

In order to be able to relate jet properties to the underlying partonic scattering, a jet algorithm that is infrared and collinear safe must be used. We will use the ATLAS default jet algorithm, i.e. the anti-k_T algorithm, with a radius parameter of 0.6.

The structure of this chapter is as follows: first we start with a discussion about the triggers used for this kind of measurement (Sect. 10.2). Section 10.3 describes the kinematic cuts and histogram binning, optimized for the trigger set-up and the geometry of the detector. QCD predictions up to next-to-leading order and the calculation of k-factors are the topic of Sects. 10.4 and 10.5 investigates theoretical uncertainties coming from the renormalization and factorization scale and the choice of PDFs. Experimental uncertainties are detailed in Sect. 10.6. Since detector effects deform the measured distribution, we need to correct the data for that. This procedure of data unfolding will be the topic of Sect. 10.7. The ATLAS sensitivity to black hole production and gravitational scattering in large extra dimensions is investigated in Sect. 10.8. Finally, conclusions are made in Sect. 10.9.

N. Boelaert, *Dijet Angular Distributions in Proton–Proton Collisions*, Springer Theses,
DOI: 10.1007/978-3-642-24597-8_10, © Springer-Verlag Berlin Heidelberg 2012

Table 10.1 Trigger menu for $10^{31}\mathrm{cm}^{-2}\mathrm{s}^{-1}$ luminosity running at $\sqrt{s} = 7\,\mathrm{TeV}$

L1 trigger	Threshold (EM scale) (GeV)	Prescale
L1_J5	5	2000
L1_J10	10	1000
L1_J20	20	50
L1_J40	40	3
L1_J70	70	5
L1_J100	100	1
L1_J130	130	1

10.2 Trigger Study

10.2.1 Trigger Menu

We will accept events passing one of the single jet triggers. See Sect. 6.2 for more details about the ATLAS trigger system.

The trigger menu that is currently expected in early data taking at $10^{31}\,\mathrm{cm}^{-2}\mathrm{s}^{-1}$ luminosity consists of the L1 single jet triggers shown in Table 10.1. All triggers follow the same naming convention, L1_JX, indicating a L1 trigger which requires that at least one jet in the event has a transverse momentum above the threshold of X GeV. For each trigger, the prescale factor is mentioned in the table as well. Prescales are used to reduce the trigger rates; for a prescale factor of *N*, only 1 in every *N* events which pass the trigger is actually accepted. In the initial stage of the LHC operation, the instantaneous luminosity is lower ($<10^{29}\,\mathrm{cm}^{-2}\mathrm{s}^{-1}$) and there is enough bandwidth available to write all events onto tape without applying a prescale. In that case, L1_J5 is used as the single jet trigger.

In this study we will use the L1_J5, L1_J40, L1_J70 and L1_J130 triggers from the expected menu and, so as to avoid bias, we will only use each trigger within the kinematic region where its efficiency curve is stable and close to 100%. In order to determine the trigger efficiency from data, the tag and probe method [2] has been used.

10.2.2 Trigger Efficiency Using the Tag and Probe Method

The tag and probe method provides an in-situ measurement of the trigger efficiency. We will use fully detector simulated dijet events generated with PYTHIA to derive the efficiencies and demonstrate a proof of principle by comparing them with the truth trigger efficiency which was used in the actual simulation of the event.

The ATLAS L1 trigger searches for so-called *Regions Of Interest* (ROIs). A ROI is a geometrical region in $(\eta,\ \phi)$ of the ATLAS detector that possibly contains interesting information about the event, and therefore needs to be further investigated at higher trigger levels. The ROIs for the triggers under consideration have an opening of (0.8, 0.8) in $(\eta,\ \phi)$.

The tag and probe method is based on matching reconstructed jets to the nearest ROI. In order to assign a fully reconstructed jet to a ROI, the matching radius R_m is used to define the distance between the jet and the center of the ROI:

$$R_m = \sqrt{\Delta\eta_{ij}^2 + \Delta\phi_{ij}^2} \tag{10.1}$$

A ROI and a jet are matched if the distance R_m between them is smaller than a given value, in this study $R_m \leq 0.4$.

The tag and probe method consists of two parts. In the first part, the *jet-level* trigger efficiency is determined. This is done in the following way:

- Run over all events in the data sample and construct a subdata sample by selecting all events passing a given trigger, e.g. the L1_J40 trigger.
- For each event in the subdata sample, match each reconstructed jet to a L1 ROI.
- For each jet in each event, check whether the matched ROI exceeds the trigger threshold. If this is the case, the jet has "passed the trigger".
- For each event in the subdata sample, "tag" randomly one of the jets for which its matched ROI passes the threshold.
- For each event in the subdata sample, disregard the tagged jet (this is to remove bias caused by the fact that we work with a triggered sample where by definition a jet is over threshold) and "probe" the remaining jet(s).
- Determine the *jet-level* trigger efficiency from the probed jets as a function of jet p_T by verifying whether the probed jets (with a certain p_T) pass the trigger which is being investigated.

The jet-level efficiency is used to determine the acceptance efficiency *P*(*A*) (*A* stands for acceptance), defined as the probability that there is a jet matching a ROI over threshold in the event under consideration:

$$P(A) = 1 - \prod P(\text{jet fail}) \tag{10.2}$$

$$= 1 - \prod (1 - P(\text{jet pass})) , \tag{10.3}$$

where the product runs over all probed jets in the event, and *P*(jet pass) is the jet-level efficiency calculated in the first step.

In the next part, the *event-level* trigger efficiency *P*(*T*) (*T* stands for trigger) is determined. This is done by looping once more over all the events in the subdata sample in order to calculate the probability that any of the jets—including the "tag" jet—satisfied the trigger in question. Mathematically, this is expressed as follows:

$$P(T) = \frac{P(A)}{P(A|T)}, \tag{10.4}$$

with $P(A|T)$ the conditional probability that there is a jet matching a ROI over threshold, given that the event triggered. This is calculated using the TrigDecisionTool class available in the ATLAS software.

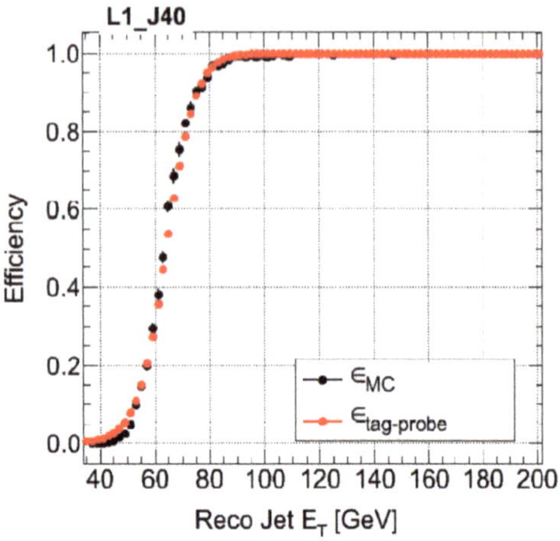

Fig. 10.1 Determination of the L1_J40 trigger efficiency ϵ obtained using the tag and probe method and Monte Carlo simulation. Figure taken from Ref. [1]

Table 10.2 Reconstructed jet p_T where the triggers become 95% efficient using both the Monte Carlo simulation and tag and probe methods

L1 trigger	L1 threshold (GeV)	Jet p_T^{reco} for which $\epsilon > 95\%$ (GeV)
L1_J5	5	>50
L1_J40	40	>80
L1_J70	70	>130
L1_J100	100	>170
L1_J130	130	>195

Using the tag and probe method, the trigger efficiencies for the triggers listed in Table 10.1 have been calculated. For example, the result from the tag and probe method for the L1_J40 trigger is shown as the red dotted curve ($\epsilon_{\text{tag}-\text{probe}}$) seen in Fig. 10.1.

Since this is a study with Monte Carlo simulated data, there is also the possibility to deduct the trigger efficiency directly by looking up the information of the event that triggered before it was digitized and sent through the detector simulation (including trigger simulation), i.e. when it comes out of the event generator; for each jet and in each event, we can see whether it passes a given trigger and we can compare that with its transverse momentum at the event generator level. Of course, this method is not applicable on real data, but in this study, it is a good test of the tag and probe method. The black dotted curve (ϵ_{MC}) in Fig. 10.1 shows the—true—trigger efficiency obtained this way. The trigger efficiency on the plateau derived using the tag and probe method agrees at the percent level with that from Monte Carlo simulation. This is also the case for the other triggers.

Table 10.2 shows for all triggers the reconstructed jet p_T where the triggers reach 95% efficiency using both the Monte Carlo simulation and the tag and probe method (with less than 1% residual difference between the two). This table will be used in the next section for deriving an optimal binning in kinematics.

10.3 Kinematic Cuts and Histogram Binning

10.3.1 Kinematic Cuts

The cuts in pseudorapidity were discussed in Chap. 4. To summarize, we repeat Eq. 4.22:

$$|\eta_1 + \eta_2| < c, \\ |\eta_1 - \eta_2| < 2\eta_{max} - c, \tag{10.5}$$

with η_1 and η_2 the pseudorapidity values of the leading and subleading jet respectively, $\eta_{\rm max}$ the maximum pseudorapidity of the two jets, dictated by the limited detector range, and c a constant. Even though ATLAS can detect jets fully up to $\eta \sim 4$, we restrict this study to jets up to $\eta_{\rm max} = 2.5$ in this study. This is because in the initial stage of ATLAS data taking (corresponding to an integrated luminosity $< 100\,{\rm pb}^{-1}$), the forward calorimeters are not well understood. Furthermore, we take $c = 1.5$, which is a trade-off between the measurable χ-range and the available statistics. We have tested that a larger value of c, e.g. $c = 2.5$, does not give a notable increase of the cross section, while it does limit the χ-range. Decreasing c, e.g. from $c = 1.5$ to $c = 0.5$, reduces the statistics too much.

With these choices, we can measure the distributions up to $\chi_{\rm max} = \exp(2\eta_{\rm max} - c) \simeq 30$.

Apart from these cuts in pseudorapidity, the dijet angular distributions are also binned in dijet invariant mass M_{jj}. The dijet invariant mass of two massless partons can be written as:

$$M_{jj} = \sqrt{p_{T1}p_{T2}}\sqrt{(\chi + 1/\chi - 2\cos(\Delta\phi))}, \tag{10.6}$$

with p_{T1} and p_{T2} the transverse momenta of the leading and subleading jet respectively, and $\Delta\phi = |\phi_1 - \phi_2|$.

Since single jet triggers will be used for this kind of study, a cut on the transverse momentum of the leading jet will be made in order to avoid bias from the trigger. Given this p_T cut and the available χ-range (going from 1 to $\chi_{\rm max}$), the minimum dijet invariant mass that can be measured without introducing a bias caused by the p_T cut, can be found by noting that, for a dijet with $p_{T1} > p_{T\rm min}$, the maximum mass possibly created is given by the LO configuration:

$$M_{jj} = p_{T,\rm min}\sqrt{(\chi_{\rm max} + 1/\chi_{\rm max} + 2)} = p_{T,\rm min}(\sqrt{\chi_{\rm max}} + 1/\sqrt{\chi_{\rm max}}) \tag{10.7}$$

Higher order configurations will always lower the p_T of the subleading jet compared to the LO configuration, so that the dijet mass obtained from higher order configurations does not exceed Eq. 10.7. Therefore, the minimum mass $M_{jj,\rm min}$ that can be measured without bias is given by Eq. 10.7.

Equation 10.7 allows us to select the dijet invariant mass bins for this study based on ATLAS trigger information; using the single jet triggers discussed in the previous

Table 10.3 Table containing for those jet triggers relevant for this study: the value of $p_{T,\min}$ corresponding to the onset of the trigger plateau increased with 10%, $M_{jj,\min}$ according to Eq. 10.7, the mass bin chosen in such a way that there is no overlap in mass bins from other triggers, and $\chi_{\max}$

Trigger	$p_{T,\min}$(GeV)	M_{jj}, min(GeV)	Mass bin (GeV)	$\chi_{\max}$
L1_J5	55	311	[320, 520]	30
L1_J40	88	500	[520, 680]	30
L1_J40	88	500	[680, 800]	30
L1_J70	143	800	[800, 1200]	30
L1_J130	215	1200	[1200, 1700]	30
L1_J130	215	1200	[1700, 2500]	30
L1_J130	215	1200	[2500, 7000]	30

section, we determine the transverse momentum where each trigger reaches 95% (see Table 10.2) and, so as to avoid problems caused by the uncertainty on the jet energy scale, we let $p_{T,\min}$ correspond to this value increased with 10%. Since we also know $\chi_{\max}$, we can determine the lower bound $M_{jj,\min}$ of the mass bin using Eq. 10.7. The upper bound $M_{jj,\max}$ is chosen in such a way that there is no overlap with mass bins from other triggers.

This is summarized in Table 10.3, which shows for the selected single jet triggers the following information: the value of $p_{T,\min}$ as discussed previously, $M_{jj,\min}$ according to Eq. 10.7, the mass bin chosen in such a way that there is no overlap in mass bins from other triggers, and $\chi_{\max}$. The trigger thresholds are from the jet trigger menu shown in Table 10.2.

10.3.2 Binning in χ

With the kinematic cuts given in Sect. 10.3.1, a binning in χ has to be defined. The main metrics used here to decide on a strategy, are *bin purity* and *bin stability*, which are obtained from comparing detector fully simulated jets with the underlying truth jets (at the generator level). These two variables quantify bin migrations between truth level spectra and reconstructed spectra.

The purity is defined as:

$$\text{Purity}_{ij} = \frac{N_{\text{rec(i)\&gen(j)}}}{N_{\text{rec(i)}}}, \tag{10.8}$$

with $N_{\text{rec(i)\&gen(j)}}$ the number of entries that were fully detector reconstructed in bin i but belong to bin j at the truth level, and $N_{\text{rec(i)}}$ the total number of reconstructed entries in bin i.

Stability is defined in a similar way as:

$$\text{Stability}_{ij} = \frac{N_{\text{rec(i)\&gen(j)}}}{N_{\text{gen(j)}}}, \tag{10.9}$$

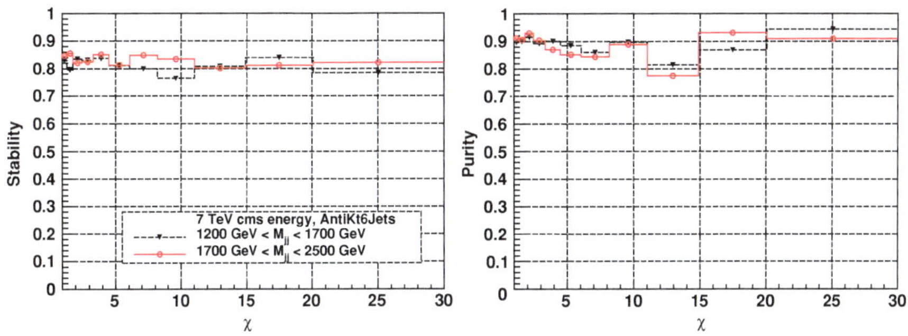

Fig. 10.2 Stability (*left* diagram) and purity (*right* diagram) for the mass bins $1200 < M_{jj} < 1700\,\text{GeV}$ and $1700 < M_{jj} < 2500\,\text{GeV}$. Figures taken from Ref. [1]

with $N_{\text{gen(j)}}$ the total number of entries in bin j at the truth level. Note that $\sum_j \text{Purity}_{\text{ij}} = 1$ for all i and $\sum_i \text{Stability}_{\text{ij}} = 1$ for all j.

An optimal binning in χ is defined as a high bin $\text{Purity}_{i=j}$ and $\text{Stability}_{i=j}$ over the whole range of χ, as then resolution effects in reconstructed data play a limited role.

Furthermore two additional criteria have been considered for obtaining an optimal binning. The first one is the ability to resolve the detailed shape of χ-spectra at low χ, where an excess of events is expected for the studied models of new physics. The second one is related to the typical cell granularity of the ATLAS hadronic calorimetry, which is 0.1 in η for the TileCal (see Sect. 6.1 for a description of the calorimeters). Small migrations of the jet axes due to the granularity can occur, which would lead to a modulation of the jet η-spectrum with a periodicity of 0.1. The same is true for observations in $|\eta_1 - \eta_2|$. In order to reduce the impact on a measurement, bin borders should be chosen where the migrations are minimal. For a measurement in χ this means that the bin edges follow an exponential curve.

It has been found that the following lower borders b_i of bins i satisfy the last two considerations:

$$b_i = e^{0.1ai}, \tag{10.10}$$

with a an integer number. Since small binsizes are desired in order to be able to resolve the shape of the distribution, a should be chosen as small as possible. However, an inferior purity and stability (values below 0.6) are obtained for bins with $a=1$ and $a=2$. But for $a=3$ continuously high values of purity and stability are obtained, as can be seen in Fig. 10.2. There is one deviation from the above formula, namely that the bin border of the highest χ bin is extended to 30 to provide the full χ range available after kinematic cuts. In summary, the binning used in the following is:

1	1.350	1.822	2.460	3.320	4.482	6.050	8.166	11.023	14.880	20.086	30

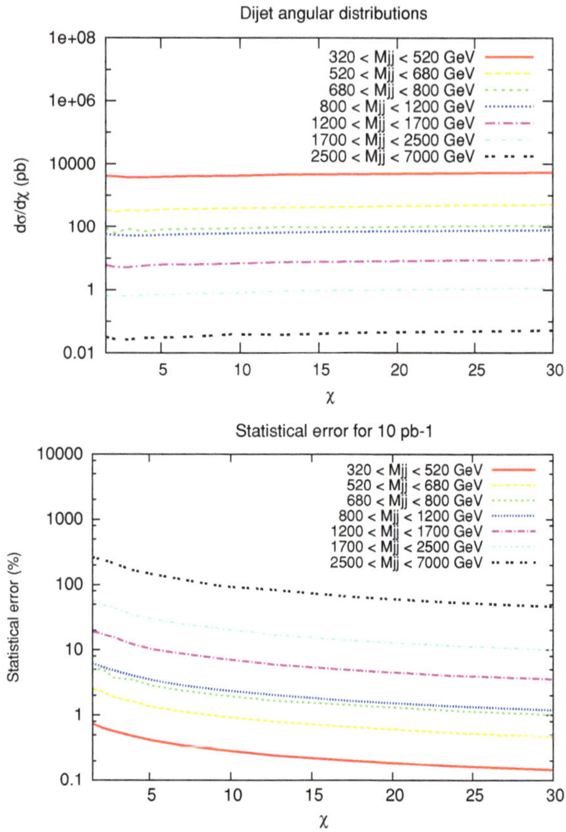

Fig. 10.3 Dijet angular distributions as calculated from QCD (*top* figure) and the statistical errors (*bottom* figure) corresponding to an integrated luminosity of $10\,\mathrm{pb}^{-1}$

10.3.3 QCD Distributions and Statistical Uncertainties for 10 pb^{-1}

Given the binning in χ discussed in Sect. 10.3.2 and the dijet mass bins proposed in Sect. 10.3.1, we show in Fig. 10.3 the QCD expectations and their statistical errors corresponding to an integrated luminosity of $10\,\mathrm{pb}^{-1}$, in the top and bottom figure respectively. Note that trigger prescales have not been taken into account for the calculation of the statistical error, since they might vary depending on the instantaneous luminosity. From the bottom figure, it is clear that the highest mass bin lacks statistics with $10\,\mathrm{pb}^{-1}$ of data, but that the other bins have reasonably sufficient statistics. When it comes to investigating new physics, we will focus on the higher mass bins, namely $1200 < M_{jj} < 1700$ GeV and $1700 < M_{jj} < 2500$ GeV.

10.4 NLO QCD Calculations and k-Factors

10.4.1 Calculating k-Factors: General Method

Monte Carlo studies investigating the behavior of ATLAS for a certain measurement, are typically carried out using so-called *complete* event generators (e.g. PYTHIA), that take a factorized approach with the major steps being the hard scattering process, initial and final state radiation, multiple interactions and non-perturbative processes such as hadronization and beam remnants. This was also discussed in Sect. 2.6. The hard scattering processes in these generators are calculated at leading order (LO) only, and even though LO calculations generally describe broad features and provide a first estimate of the cross sections, next-to-leading (NLO) calculations are a must in order to get more precise estimates (see Chap. 3).

In order to compensate for that, we can multiply fully perturbative LO calculations bin-by-bin with the so-called k-factors that take into account perturbative NLO corrections. The k-factors have been derived from QCD calculations using NLOJET++ [3] and PYTHIA [4].

As stated before, PYTHIA is a complete event generator, while NLOJET++ only calculates the hard scattering—using NLO matrix elements—and none of the other steps, which makes the generator incomplete.

Since the parton showers in PYTHIA take over part of the NLO corrections, care needs to be taken when combining the two generators for the derivation of the k-factors.

Assume a distribution in x for which a *complete* LO Monte Carlo sample (including hard scattering, showers and non-perturbative effects) has been generated: $\mathbf{run_{COMPL}}(x)$. For the derivation of the k-factor (as function of x), three different types of Monte Carlo calculations are needed:

1. $\mathbf{run_{LO}}(x)$: LO matrix elements: done with PYTHIA, run_{LO} (PYTHIA)(x), and NLOJET++, run_{LO} (NLOJET++)(x)
2. $\mathbf{run_{NLO}}(x)$: NLO matrix elements with NLOJET++
3. $\mathbf{run_{SHOW}}(x)$: LO PYTHIA with parton showers (PSs) only and non-perturbative processes 'off'

These runs should be calculated with the same parameter settings and selection cuts, apart from the choice of PDFs. To ensure consistency, a LO PDF is needed for the LO calculations, while the NLO calculation needs an NLO PDF.

Having all these runs, the accuracy of $\mathbf{run_{COMPL}}(x)$ can be improved using NLO knowledge in the following way:

$$\text{run}_{\text{COMPL}}(x) \times \frac{\left(\text{run}_{\text{NLO}}(x)/\text{run}_{\text{LO}}(\text{NLOJET++})(x)\right)}{\left(\text{run}_{\text{SHOW}}(x)/\text{run}_{\text{LO}}(\text{PYTHIA})(x)\right)} = \text{run}_{\text{COMPL}}(x) \times \frac{\text{run}_{\text{NLO}}(x)}{\text{run}_{\text{SHOW}}(x)}, \tag{10.11}$$

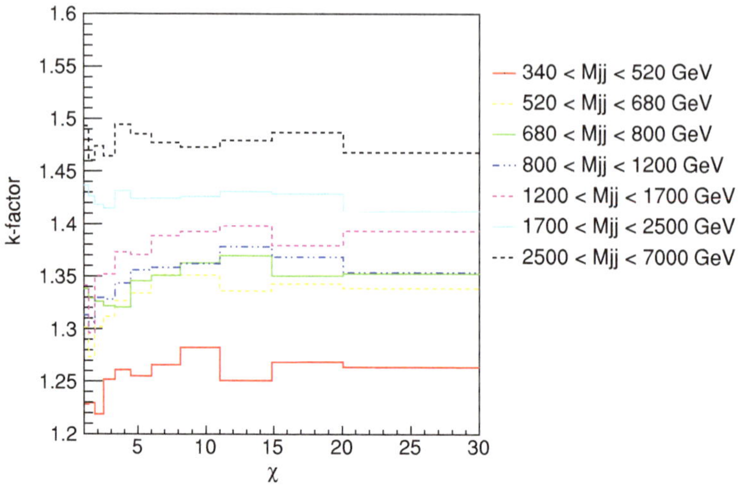

Fig. 10.4 K-factors for the dijet angular distributions in the dijet mass bins that were defined in Sect. 10.3

where the equals sign only holds for run_{LO} (NLOJET++) $\equiv$ run_{LO} (PYTHIA). Possible differences in the parametrization of the strong coupling constant make this a non-trivial assumption.

From Eq. 10.11, the k-factor can be defined as:

$$k(x) = \frac{\left(\text{run}_{\text{NLO}}(x)/\text{run}_{\text{LO}}(\text{NLOJET}++)(x)\right)}{\left(\text{run}_{\text{SHOW}}(x)/\text{run}_{\text{LO}}(\text{PYTHIA})(x)\right)} = \frac{\text{run}_{\text{NLO}}(x)}{\text{run}_{\text{SHOW}}(x)}, \tag{10.12}$$

which can then be used to scale $\mathbf{run}_{\mathbf{COMPL}}(x)$. Again the last equal sign only holds for run_{LO} (NLOJET++) $\equiv$ run_{LO} (PYTHIA).

10.4.2 NLO QCD Calculations and k-Factors for the Dijet Angular Distributions

Figure 10.4 shows the k-factors for the dijet angular distributions in the dijet mass bins that were defined in Sect. 10.3. The LO calculations have been done with the CTEQ6L1 PDF [5], while the NLO calculations have been done with the CTEQ66 PDF [6]. From the figure, it can be noticed that the k-factors are rather flat and close to unity, which indicates that higher order corrections are at a controllable level. The k-factors vary between 1.2 and 1.55 and are closest to unity for the lowest mass bin, while the largest deviations show up for the highest bin.

Since most of the early new physics searches will be done in the mass bins $1200 < M_{jj} < 1700$ GeV and $1700 < M_{jj} < 2500$ GeV, the different runs used for

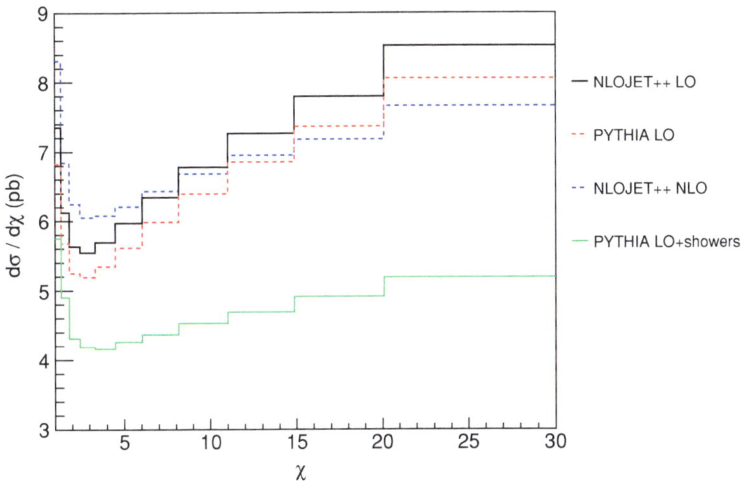

Fig. 10.5 The different runs used for the calculation of the k-factor (Eq. 10.12) for the dijet angular distribution for the mass bin $1200 < M_{jj} < 1700$ GeV. See the text for more information about the runs

the k-factor calculation in these bins, are shown in Figs. 10.5 and 10.6 respectively. From these figures, it can be seen that the LO cross sections calculated with NLOJET++ are about 4–5% higher than the ones calculated with PYTHIA. This is because the programs use different values of $\alpha_s(M_z^2)$: 0.129783 in NLOJET++ and 0.1265 in PYTHIA. Since $(0.129783/0.1265)^2 \approx 1.0526$, the LO cross section obtained with NLOJET++ is indeed about 5% higher than the PYTHIA one. In fact, for the mass bins under study the ratio will be a bit lower because of the running of α_s. For example for a 200 GeV jet and a LO running of α_s (see Eq. 2.11), the ratio becomes $(1.0811/1.0582)^2 \approx 1.042$.

Previous studies carried out at the Tevatron, do not always mention the use of k-factors. Instead, their—completely equivalent—approach was to correct a perturbative NLO QCD calculation with non-pertubative effects such as hadronization and the underlying event [7, 8]. These corrections are determined on a bin-by-bin basis by the ratio of a complete LO sample and a sample with non-perturbative effects and the underlying event turned off [8], which is, with the notation used here, $\text{run}_{\text{COMPL}}(x)/\text{run}_{\text{SHOW}}(x)$. The end result equals $\text{run}_{\text{NLO}}(x)\text{run}_{\text{COMPL}}(x)/\text{run}_{\text{SHOW}}(x)$, which is Eq. 10.11, apart from possible differences between run_{LO} (NLOJET++) and run_{LO} (PYTHIA).

Both approaches are equivalent when making QCD predictions, but sometimes k-factors are also used for new physics distributions. This was done in the past [7] and the ATLAS collaboration has opted for the same approach.

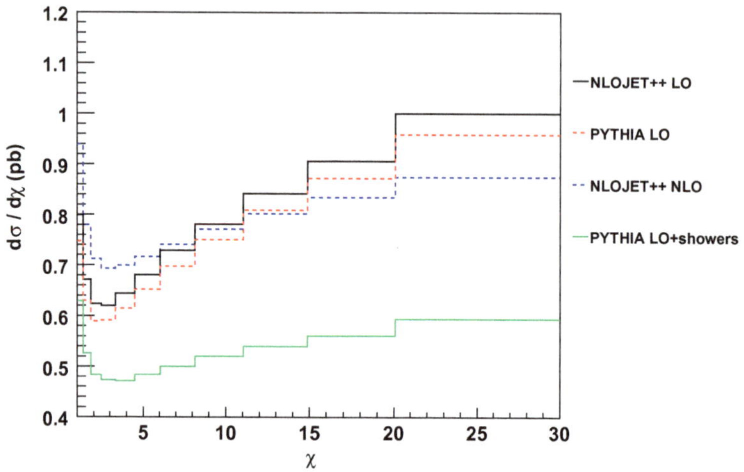

Fig. 10.6 The different runs used for the calculation of the k-factor (Eq. 10.12) for the dijet angular distribution for the mass bin $1700 < M_{jj} < 2500\,\text{GeV}$. See the text for more information about the runs

Beyond the Standard Model physics (BSM) causes an increase in the cross section:

$$\frac{d\sigma_{\text{total}}}{d\chi} = \frac{d\sigma_{\text{QCD}}}{d\chi} + \frac{d\sigma_{\text{BSM}}}{d\chi} \tag{10.13}$$

Since k-factors are typically > 1, which can be seen from Fig. 10.4 and was also reported by Ref. [7], this means that multiplying the new physics distributions with k-factors will move them away from the QCD curves, making it more difficult to discover new physics when making a comparison with data. Instead, one can put tighter limits on new physics models this way.

10.5 Theoretical Uncertainties

10.5.1 Renormalization and Factorization Scale Uncertainties

The uncertainty of the renormalization (μ_R) and factorization (μ_F) scales contributes to a systematic error. We have studied this by letting μ_R and μ_F vary independently between 0.5, 1 and 2 times the average transverse momentum μ_0 of the dijet ($\mu_0 = (p_{T1} + p_{T2})/2$), resulting in nine different distributions in total. The central member has $\mu_R = \mu_F = \mu_0$, and all calculations have been done at NLO using NLOJET++.

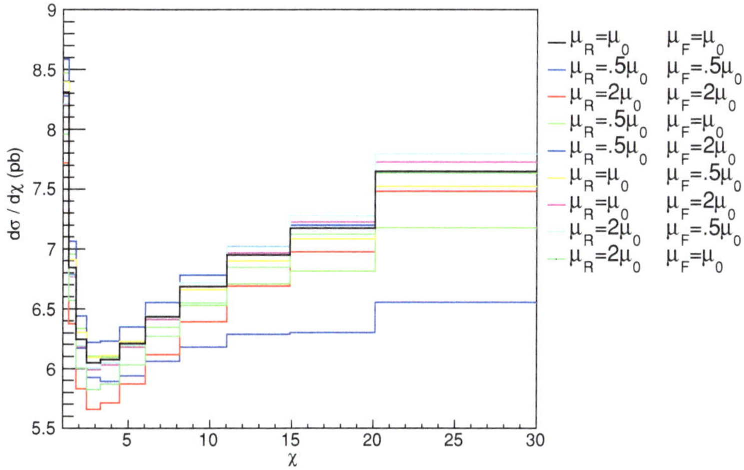

Fig. 10.7 Dijet angular distributions for the mass bin $1200 < M_{jj} < 1700$ GeV, and for different choices of μ_R and μ_F

The scale uncertainties have been calculated both for the distributions normalized to unit area $1 < \chi < 30$ and unnormalized. Each scale uncertainty is split up in a positive (+) uncertainty and a negative (-) uncertainty, defined around the central member:

$$+\text{scale uncertainty}(\chi) = \frac{\max_{\{r,f\}=\{0.5,1.,2.\}}\left(\mathrm{d}\sigma/\mathrm{d}\chi_{(\mu_R=r\mu_0,\mu_F=f\mu_0)} - \mathrm{d}\sigma/\mathrm{d}\chi_{(\mu_R=\mu_F=\mu_0)}\right)}{\mathrm{d}\sigma/\mathrm{d}\chi_{(\mu_R=\mu_F=\mu_0)}}$$

$$-\text{scale uncertainty}(\chi) = \frac{\max_{\{r,f\}=\{0.5,1.,2.\}}\left(\mathrm{d}\sigma/\mathrm{d}\chi_{(\mu_R=\mu_F=\mu_0)} - \mathrm{d}\sigma/\mathrm{d}\chi_{(\mu_R=r\mu_0,\mu_F=f\mu_0)}\right)}{\mathrm{d}\sigma/\mathrm{d}\chi_{(\mu_R=\mu_F=\mu_0)}} \quad (10.14)$$

Figure 10.7 shows all nine distributions for the mass bin $1200 < M_{jj} < 1700$ GeV. Figure 10.8 shows the same distributions but normalized to unit area under $1 < \chi < 30$.

Using Eq. 10.14, we calculate the positive and negative uncertainties for the mass bins defined in Table 10.3. This is shown in Fig. 10.9. The procedure is repeated for the angular distributions normalized to unit area, and the results are shown in Fig. 10.10. Normalizing the distributions reduces the uncertainties somewhat.

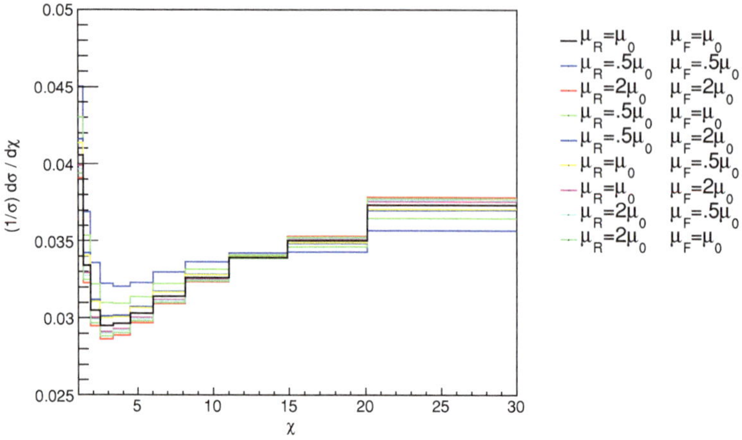

Fig. 10.8 Dijet angular distributions normalized to unit area $1 < \chi < 30$, for the mass bin $1200 < M_{jj} < 1700\,\text{GeV}$, and for different choices of μ_R and μ_F

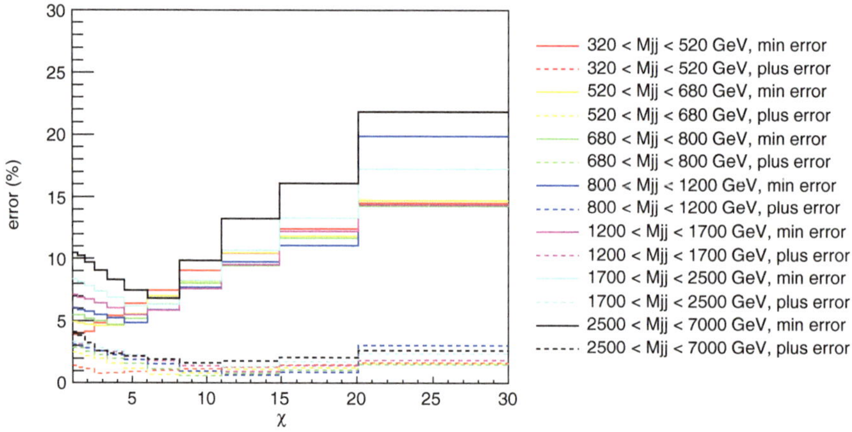

Fig. 10.9 Systematic uncertainty coming from the choice of $\mu_{R,F}$ for the dijet angular distributions. The *full* lines denote the *positive* uncertainties, while the *dashed* lines denote the *negative* uncertainties (see Eq. 10.14)

10.5.2 PDF Uncertainties

We have calculated the angular distributions for all 44 error members of the CTEQ66 PDF and applied the Master Equation suggested in Ref. [9] to deduct a positive and negative uncertainty on a quantity X:

$$\Delta X^+_{\max} = \sqrt{\sum_{i=1}^{N} \left[\max\left(X_i^+ - X_0, X_i^- - X_0, 0 \right) \right]^2} \tag{10.15}$$

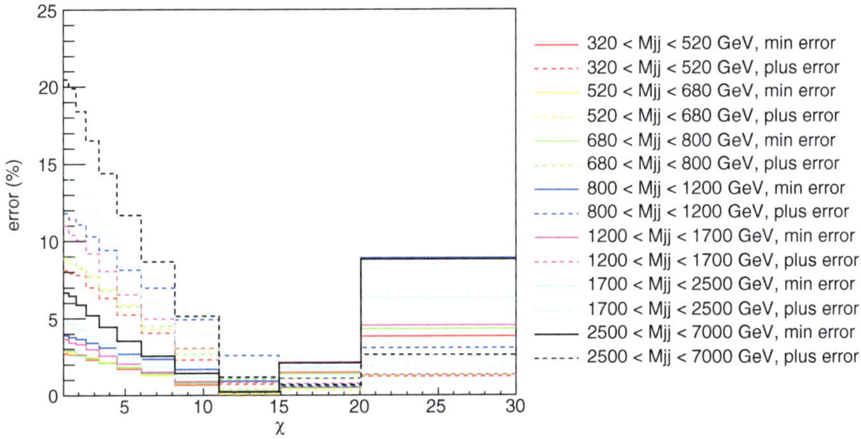

Fig. 10.10 Systematic uncertainty coming from the choice of $\mu_{R,F}$, for the dijet angular distributions normalized to unit area $1 < \chi < 30$. The *full* lines denote the *positive* uncertainties, while the *dashed* lines denote the *negative* uncertainties (see Eq. 10.14)

$$\Delta X^{-}_{\max} = \sqrt{\sum_{i=1}^{N} \left[\max\left(X_0 - X_i^{+}, X_0 - X_i^{-}, 0 \right) \right]^2}. \tag{10.16}$$

ΔX^{+} adds in quadrature the PDF error contributions that lead to an increase in the observable X, and ΔX^{-} the PDF error contributions that lead to a decrease.

In Fig. 10.11 we plot the positive and negative uncertainties on the dijet angular distributions for all mass bins. The uncertainties are largest for the highest mass bin. This is not surprising, given the lack of experimental data for such high masses to constrain the gluon PDF. Fortunately, the uncertainties in PDFs mainly cause an uncertainty in absolute normalization of the distributions, rather than in shape. This is confirmed by Fig. 10.12 which shows the PDF uncertainties on the distributions normalized to unit area $1 < \chi < uj$ 30. Apart from the first χ bins, the error is less than 1%.

10.6 Experimental Uncertainties

10.6.1 General Considerations

The measurement of dijet angular distributions uses the kinematics of the two leading reconstructed jets. Any uncertainty in these quantities will give rise to a systematic

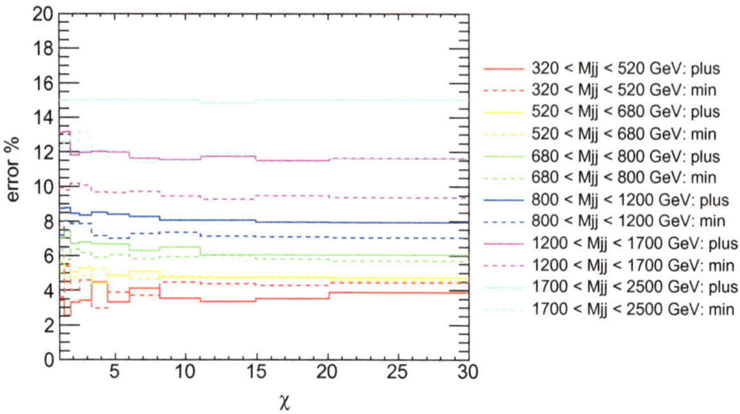

Fig. 10.11 PDF uncertainties for the dijet angular distributions. The *full* lines denote the *positive* uncertainties, while the *dashed* lines denote the *negative* uncertainties (see Eq. 10.16)

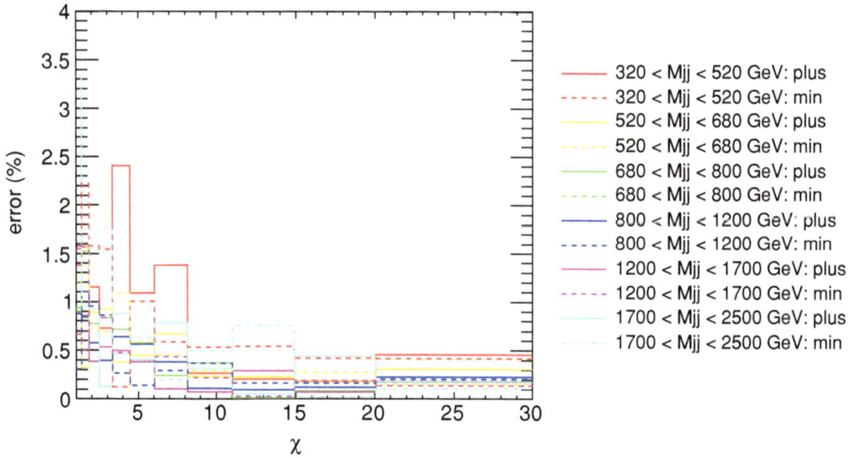

Fig. 10.12 PDF uncertainties for the dijet angular distributions normalized to unit area $1 < \chi < 30$. The *full* lines denote the *positive* uncertainties, while the *dashed* lines denote the *negative* uncertainties (see Eq. 10.16)

uncertainty in the measurement. The major effects come from uncertainties on the reconstructed energy, pseudorapidity and azimuthal angle.

Since the ATLAS calorimeters are non-compensating, a set of energy calibrations needs to be applied to the reconstructed jets before their kinematics correspond to the hadronic level. The various calibration methods were discussed in Chap. 7. Having to apply these correction procedures will cause systematic errors on a measurement, which can be divided up into two categories, namely *Jet Energy Scale (JES)* uncertainties and *Jet Energy Resolution (JER)* uncertainties. JES uncertainties can be further split up in *Absolute JES* and *Relative JES* uncertainties.

The Absolute JES uncertainty can be thought of as a shift in the mean of the fractional difference in the energy between the reconstructed calorimeter jet and its associated truth jet. This shift affects all jets and can originate from biases introduced by the calibration methods (limited statistics and intrinsic biases for the in-situ methods, disagreements between Monte Carlo predictions for Monte Carlo based methods, etc.,). A conservative number is 10% [10].

The Relative JES uncertainty represents the variation in the uniformity in the energy scale across different detector regions. Such shifts introduce a pseudorapidity dependency of the Jet Energy Scale, and can be due to different detector technologies, different geometrical layouts of the detector components, presence of local dead cells, etc. ATLAS uses the balance of dijet events to improve the relative scale (see also Sect. 7.4.4), and a conservative uncertainty is 3% [10].

The resolution is given by the width of the distribution of the relative difference between the reconstructed energy of the jet and its associated truth jet. In Ref. [11], it was shown that the resolution can be parametrized as:

$$\frac{\sigma_E}{E} = \frac{A}{E} \oplus \frac{B}{\sqrt{E}} \oplus C \tag{10.17}$$

where A is the noise term (electronic noise, important at low energies), B is the stochastic term (Poissonian event-to-event fluctuations, for instance in the energy deposited in the active calorimeter volume or in the particle composition of the jet), and C is the constant term (effects proportional to the jet energy, like non-uniformity in the calorimeter structure, or the presence of cracks and dead material). The values ($A = 4.9 \pm 0.2$, $B = 0.62 \pm 0.03$, $C = 0.027 \pm 0.002$) were derived using the de dijet balance technique (see also Sect. 7.4.4).

Apart from JES uncertainties and a finite energy resolution, also uncertainties in pseudorapidity and azimuthal angle must be considered. In Sect. 7.3, we saw that the input for a jet algorithm has the typical size of a tower, $\Delta\eta \times \Delta\phi = 0.1 \times 0.1$. Hence, a conservative estimate for uncertainties in pseudorapidity and azimuthal angle is of that order. Note that the size of the actual calorimeter cells can be much smaller (depending on their location in the detector).

10.6.2 Estimate of Experimental Uncertainties

We have used PYTHIA in order to investigate the importance of the different aspects of experimental uncertainties.

First we have generated dijet events in a reference sample, meaning that we have not applied any disturbance. Next we have generated a few samples aimed at imitating the detector effects discussed in the previous section:

- Variation of the Absolute JES (AJES): we have increased and decreased the jet energy with 10% respectively.

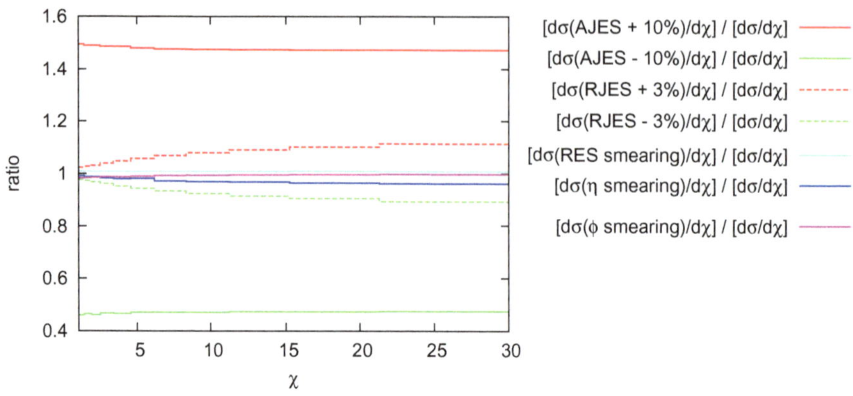

Fig. 10.13 Ratio of modified dijet angular distributions over reference cross section, in the mass bin $1700 < M_{jj} < 2500$ GeV. The cross sections are modified in order to represent the effects coming from the absolute and relative jet energy scale (AJES and RJES) uncertainties, a finite resolution (RES) and angular smearing in pseudorapidity and azimuthal angle

- Variation of the Relative JES (RJES): we have varied the jet energy linearly in η, with 0% at $\eta = 0$, and 3 or -3% at $|\eta| = 3$. This choice is motivated by the fact that the reference scale is at $\eta = 0$.
- Investigation of the jet energy resolution (JER): we have varied the jet energy according to a normal distribution with standard deviation given by the resolution in Eq. 10.17.
- Investigation of the polar angle: we have smeared the pseudorapidity of each jet according to a normal distribution with zero mean and standard deviation 0.1.
- Investigation of the azimuthal angle: we have smeared the azimuthal angle, again using a normal distribution with zero mean and standard deviation 0.1.

We have used these samples to construct the dijet angular distribution in the mass bin $1700 < M_{jj} < 2500$ GeV (with the binning in χ as defined in Sect. 10.3.2). Figure 10.13 shows the resulting distributions normalized to the reference distribution. The smearing in pseudorapidity and azimuthal angle does not cause a large deviation from unity, nor do resolution effects.

However, a change in the jet energy does have a large impact. This is caused by the fact that the angular distributions are binned in dijet invariant mass. Since the dijet invariant mass spectrum is a steeply falling distribution, a shift in dijet invariant mass (caused by a shift in the jet energy scale) will have a large impact on the dijet angular distribution.

Compared to the changes in absolute energy scale, the changes in the relative energy scale are modest in size, since the Relative JES is believed to be much more controllable than the Absolute JES (3 versus 10%). However, an absolute shift in jet energy will mainly influence the normalization of the dijet angular distributions. But the non-uniform shift in jet energy will cause variations both in shape and normalization.

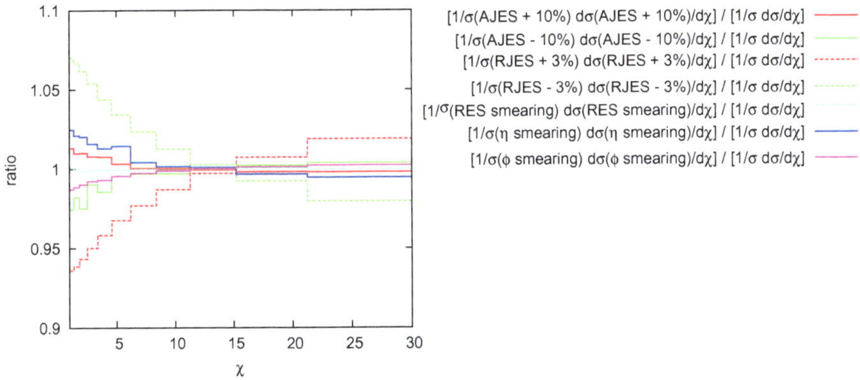

Fig. 10.14 Ratio of modified dijet angular distributions over reference cross section, in the mass bin $1700 < M_{jj} < 2500$GeV. The dijet angular distributions are normalized to unit area $1 < \chi < 30$. The cross sections are modified in order to represent the effects coming from the absolute and relative jet energy scale (AJES and RJES) uncertainties, a finite resolution (RES) and angular smearing in pseudorapidity and azimuthal angle

The uncertainty coming from the Absolute JES can be largely reduced by normalizing the dijet angular distributions to unit area $1 < \chi < 30$. This can be seen in Fig. 10.14, which shows the uncertainties for the distributions normalized to unit area. With the exception of the Relative JES uncertainty, the effects from all other uncertainties have been largely reduced.

The same conclusions hold for the other mass bins. To illustrate this, we show the uncertainties for the normalized dijet angular distributions in the mass bin $320 < M_{jj} < 520$ GeV in Fig. 10.15. In this bin, the systematic uncertainty on the Relative JES has less impact than in the higher bin.

10.7 Data Unfolding

Distributions measured by ATLAS are usually deformed by detector effects. To illustrate this, we use Monte Carlo data to compare a fully detector simulated and reconstructed "measurement" (typically referred to as RECO) of the dijet angular distributions, with its true physics distribution (TRUTH). Figure 10.16 shows the ratio of RECO to TRUTH for the mass bins defined earlier. The ratio is very close to unity, with deviations not more than a few percent. This is not surprising since the binning in χ was optimized to reduce bin migration (see Sect. 10.3.2).

A measured distribution can be seen as a convolution of the true physics distribution with the detector response. Therefore, in order to be able to study physics, we need to *unfold* the measurement, so that the result corresponds to a physical distribution. A *migration matrix* is used to describe the migration between true

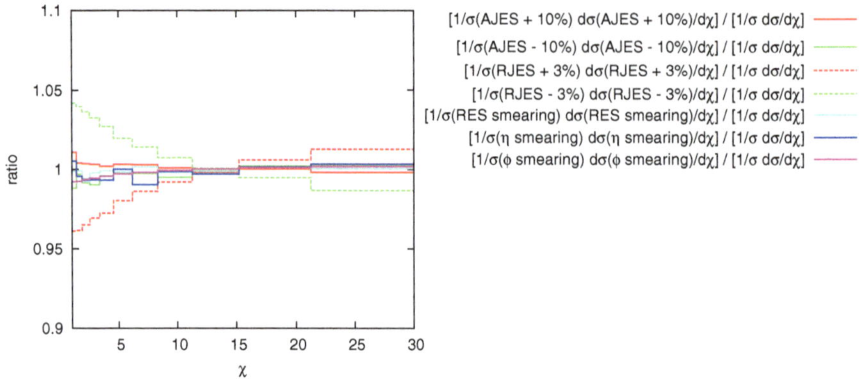

Fig. 10.15 Ratio of modified dijet angular distributions over reference cross section, in the mass bin $320 < M_{jj} < 520$GeV. The dijet angular distributions are normalized to unit area $1 < \chi < 30$. The cross sections are modified in order to represent the effects coming from the absolute and relative jet energy scale (AJES and RJES) uncertainties, a finite resolution (RES) and angular smearing in pseudorapidity and azimuthal angle

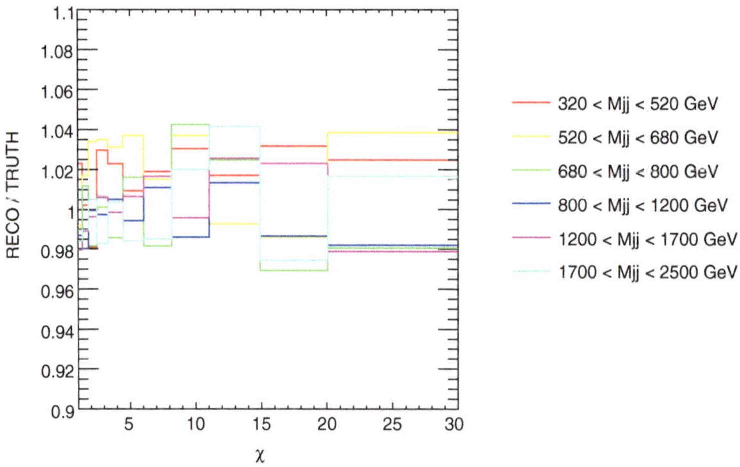

Fig. 10.16 Ratio of the dijet angular distributions that are fully detector simulated an reconstructed (REC0) over the true physics distribution (TRUTH)

physics bins and the measurement. For the dijet angular distributions, we will have one migration matrix for each dijet mass bin.

We can find this matrix by looking at Monte Carlo data and comparing the RECO data with the underlying truth events. For each event we fill a 2-dimensional histogram with $(\chi_{\text{reco}}, \chi_{\text{truth}}, w)$, with w the event weight, being the cross section times the integrated luminosity. Figure 10.17 shows the migration matrix as a scatter plot for the dijet angular distribution in the mass bin $1700 < M_{jj} < 2500$ GeV, corresponding to an integrated luminosity of 1 pb^{-1}.

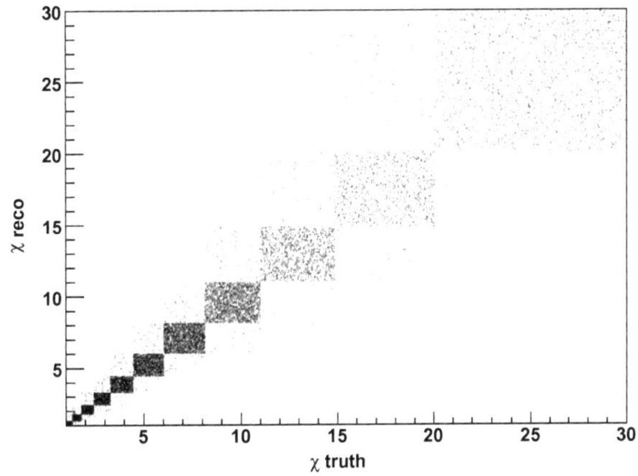

Fig. 10.17 Scatter plot of $(\chi_{\text{reco}}, \chi_{\text{truth}}, w)$, also known as the migration matrix, for the dijet angular distribution in the mass bin $1700 < M_{jj} < 2500\text{GeV}$

Given the migration matrix, we can recover the true physics distribution, by taking the inverse and multiplying this with the measured distribution. Since the migration matrix is typically ill behaved, algebraic techniques such as Singular Value Decomposition are often used to get more accurate results [12]. Because the ratio of RECO over TRUTH is close to unity, we expect to be able to perform a data unfolding rather well.

10.8 Sensitivity to Black Hole Production and Gravitational Scattering in Large Extra Dimensions

This section concentrates on GravADD processes [gravitational scattering (GS) and semi-classical black holes (BH)] with early data, with an integrated luminosity ranging from 1 to 10 pb^{-1}. We study the dijet angular distributions normalized to unit area $1 < \chi < 30$, in the mass bins $1200 < M_{jj} < 1700$ GeV and $1700 < M_{jj} < 2500$ GeV. We fix the number of extra dimensions to $n = 6$. We have checked that changing the number of extra dimensions (e.g. from 4 to 6) does not modify the distributions drastically. However, the parameters M_{eff} and M_s/M_P (see Sect. 5.3 for a description of the parameters) do have a large influence on the phenomenology.

The approach is similar to what was done in Sect. 9.4 and which is explained in detail in appendix A. We calculate $(\mathrm{d}N_{\text{QCD}}/\mathrm{d}\chi)/N_{\text{QCD}}$ and $(\mathrm{d}N_{\text{total}}/\mathrm{d}\chi)/N_{\text{total}}$, with N_{QCD} and N_{total} respectively the number of QCD and total (= QCD + GS + BH) events corresponding to a certain integrated luminosity. We then perform a chi-square (χ^2) test between these distributions to test the null hypothesis that $(\mathrm{d}N_{\text{total}}/\mathrm{d}\chi)/N_{\text{total}}$ follows the QCD distribution. We use both a statistical and a systematic uncertainty for the calculation of χ^2:

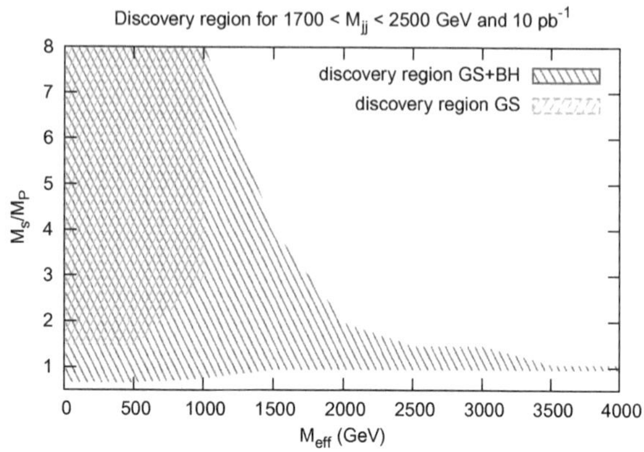

Fig. 10.18 Discovery region for the mass bin $1700 < M_{jj} < 2500$ GeV at $10\,\text{pb}^{-1}$, assuming a 25% systematic uncertainty. The *green* region is the discovery region for GS only, the *blue* area is for BH production and GS together

$$\chi^2 = \sum_{\text{allbins}i} \left[\frac{\left(\frac{N_{\text{QCD},i}}{N_{\text{QCD}}} - \frac{N_{\text{total},i}}{N_{\text{total}}} \right)^2}{s^2_{\text{stat},i} + s^2_{\text{sys},i}} \right], \tag{10.18}$$

where $N_{\text{QCD},i}$ and $N_{\text{total},i}$ are the number of QCD and total events respectively in bin i. The statistical error $s_{\text{stat},i}$ is for each bin taken as $\sqrt{N_{\text{QCD},i}}$, and the systematic error $s_{\text{sys},i}$ is taken to be 25% over the whole χ range.

Since the distributions are normalized to unit area, the quadratic sum of all experimental uncertainties does not exceed 10% (see e.g. Fig. 10.14), while the theoretical uncertainties—dominated by the renormalization scale uncertainty—remain under the 15% (see Sect. 10.5). Hence, the assumption of a 25% total systematic uncertainty is a conservative one.

Using χ^2, we can calculate the p-value:

$$p = \int_{\chi^2}^{\infty} f(z; n_d)\text{d}z, \tag{10.19}$$

with $f(z; n_d)$ the probability density function of the chi-square statistic with $n_d =$ (number of bins -1) degrees of freedom. The p-value can be understood as the probability, under the assumption of the null hypothesis, of obtaining data at least as incompatible with the null hypothesis as is actually observed. Values for p can be found in standard math references. The null hypothesis of identity is rejected for $p < 1.35 \times 10^{-3}$, corresponding to a significance $Z = 3$ (see Appendix A), and the parameter region for which this occurs, is denoted as discovery region.

Figure 10.18 shows the discovery region obtained for the dijet angular distribution in the mass bin $1700 < M_{jj} < 2500$ GeV and with $10\,\text{pb}^{-1}$ integrated luminosity. The smaller green area is the discovery area for gravitational scattering as new physics process only, while the blue area results when one includes black holes as well. Note that small values of M_s/M_P ($M_s/M_P < 1$) fall outside the sensitivity region. The reason is the absence of black hole creation because in that region the lower limit on

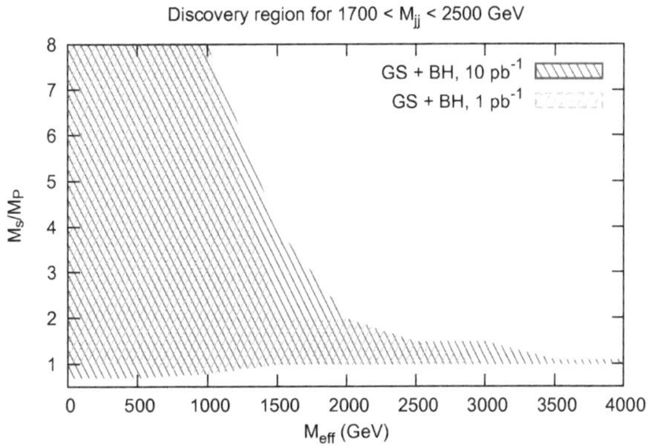

Fig. 10.19 Discovery region for the mass bin $1700 < M_{jj} < 2500$GeV for GS and BH production for an integrated luminosity of $1\,\text{pb}^{-1}$ (*green* lines) and $10\,\text{pb}^{-1}$ (*blue* lines), assuming a 25% systematic uncertainty

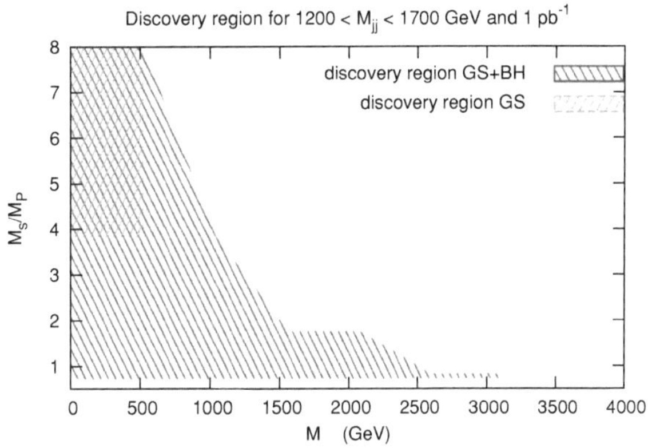

Fig. 10.20 Discovery region for the mass bin $1200 < M_{jj} < 1700$ GeV at $1\,\text{pb}^{-1}$, assuming a 25% systematic uncertainty. The *green* region is the discovery region for GS only, the *blue* area is for BH production and GS together

the black hole mass is drastically increasing with decreasing M_s/M_P. See Eq. 5.40 and the discussion underneath.

For an integrated luminosity of $1\,\text{pb}^{-1}$, we expect a lot less statistics in this mass bin (see Fig. 10.3 for a plot with the expected statistics). In Fig. 10.19 we compare the discovery region for GS and BH production for an integrated luminosity of 1 and $10\,\text{pb}^{-1}$, in the mass bin $1700 < M_{jj} < 2500$ GeV. The discovery region reduces, but not that much since the difference between new physics and QCD overcomes the statistical uncertainty.

The statistics in the lower mass bin $1200 < M_{jj} < 1700$GeV is better, but this cannot compensate the drop of new physics signal due to the lower masses being considered. Figure 10.20 shows the discovery region both for BH production and GS, and for GS only, for the mass bin $1200 < M_{jj} < 1700$ and with an integrated luminosity of $1\,\text{pb}^{-1}$. Compared to the higher mass bin (Fig. 10.19), the discovery region is smaller.

The conclusion is that, for an integrated luminosity ranging from 1 or $10\,\text{pb}^{-1}$, we are most sensitive to GravADD processes in the mass bin $1700 < M_{jj} < 2500$GeV. Masses above 2,500 GeV have not been considered since this study focuses on an early data measurement and the jet calibration for high p_T jets will initially suffer from large uncertainties.

10.9 Conclusions

This chapter has investigated the various aspects that go into the measurement of dijet angular distributions with ATLAS.

We have used the trigger menu and the detector layout to define the selection cuts in pseudorapidity and to optimize the binning in dijet invariant mass and χ. Migration effects in χ are at a minimal level when using a binning that increases exponentially with χ. The mass bins have been chosen in such a way that the measurement does not rely on trigger turn-on effects or jet energy scale uncertainties, but instead makes optimal use of the plateau where the trigger reaches its maximum efficiency.

The dijet angular distributions have been calculated up to NLO in the case of QCD, and k-factors have been derived. The k-factors are rather flat and in the neighborhood of unity, which indicates that higher order corrections are at a controllable level.

For each mass bin, we have quantified the theoretical and experimental uncertainties on the distributions, assuming a realistic but conservative performance of the jet reconstruction. Most of the uncertainties can be reduced by normalizing the distributions. After normalization, the theoretical uncertainty is dominated by the choice of the scales used in the calculation, while a non-uniform (pseudorapidity) dependent uncertainty of the Jet Energy Scale is the most important uncertainty at the experimental site.

Using Monte Carlo data, we have discussed unfolding issues. Since the ratio of a fully detector simulated and reconstructed distribution over its true physical distribution is close to unity, we expect to be able to perform the data unfolding rather well. We have shown the migration matrix for a particular mass bin and noticed that the migrations between different χ bins are rather small.

The sensitivity of ATLAS to gravitational effects coming from large extra dimensions has been investigated, and it turns out that already with 1 pb^{-1}, the ATLAS experiment is sensitive to these effects in the high mass bins. For a luminosity ranging from 1 or 10 pb^{-1}, ATLAS is most sensitive to GravADD processes in the—normalized—mass bin $1700 < M_{jj} < 2500$.

References

1. N. Boelaert, G. Choudalakis, P.O. Deviveiros, E. Feng, H. Li, J. Poveda, L. Pribyl, F. Ruehr, S.L. Wu, *ATLAS Sensitivity to Contact Interactions and Gravity Mediated Effects in Large Extra Dimensions Using Dijet Events at* $\sqrt{s} = 7$TeV, Technical Report ATL-COM-PHYS-2010-136, (CERN, Geneva, 2010)
2. T. LeCompte, *Level 1 Jet Efficiency Measurements Using the Tag and Probe Method*, Technical Report ATL-COM-PHYS-2007-014, (CERN, Geneva, 2007)

3. Z. Nagy, Next-to-leading order calculation of three-jet observables in hadron-hadron collision. Phys. Rev. **D68**, 094002 (2003). (hep-ph/0307268)
4. T. Sjöstrand, S. Mrenna, P. Skands, A brief introduction to PYTHIA 8.1. JHEP **05**, 026 (2007). (hep-ph/07103820)
5. D. Stump, J. Huston, J. Pumplin, W.K. Tung, H.L. Lai, S. Kuhlmann, J.F. Owens, Inclusive jet production, parton distributions, and the search for new physics. JHEP **0310**, 046 (2003). (hep-ph/0303013)
6. P.M. Nadolsky, H.L. Lai, Q.-H. Cao, J. Huston, J. Pumplin, D. Stump, W.K. Tung, C.P. Yuan, Implications of CTEQ global analysis for collider observables. Phys. Rev. **D78**, 013004 (2008)
7. D0 Collaboration, Measurement of dijet angular distributions at $\sqrt{s} = 1.96$TeV and searches for quark compositeness and extra spatial dimensions. Phys. Rev. Lett. **103**, 191803 (2009). (hep-ex/09064819)
8. CDF Collaboration, T. Aaltonen et al., Search for new particles decaying into dijets in proton-antiproton collisions at $\sqrt{s} = 1.96$TeV. Phys. Rev. D **79**, 112002 (2009). (hep-ex/08124036)
9. J.M. Campbell, J.W. Huston, W.J. Stirling, Hard interactions of quarks and gluons: a primer for LHC physics. Rep. Prog. Phys. **70**, 89–193 (2007). (hep-ph/0611148)
10. ATLAS Collaboration, *Expected Performance of the ATLAS Experiment: Detector, Trigger and Physics*, (CERN, Geneva, 2008)
11. R. Piegaia, G. Romeo, A. Schwartzman, *Measurement of the Jet Energy Resolution in ATLAS using Data-Driven Techniques*, Technical Report ATL-PHYS-INT- 2010-029, (CERN, Geneva, 2010)
12. A. Hoecker, V. Kartvelishvili, SVD approach to data unfolding. Nucl. Instrum. Meth. A **372**, 469 (1996). (hep-ph/9509307)

Chapter 11
Measurement of Dijet Angular Distributions by ATLAS

11.1 Introduction

This chapter presents the first measurement of dijet angular distributions in *pp* collisions at $\sqrt{s} = 7$ TeV by the ATLAS detector. The dataset consists of about 31 million events that were recorded in 2010 during the months April and May. The dijet angular distributions have been measured in the mass bins $340 < M_{jj} < 520$ GeV and $520 < M_{jj} < 680$ GeV, and have been compared to Monte Carlo simulations.

Rather than using pseudorapidity, rapidity is used to define the angular variable χ:

$$\chi = e^{|y_1 - y_2|} \tag{11.1}$$

A motivation for this change will be given in Sect. 11.4. The distributions are measured up to $\chi = 30$, using the same binning as outlined in Sect. 10.3.2. In order to reduce the systematic uncertainties, the distributions are normalized to unit area.

11.2 Data Selection

Run selection criteria have been discussed in Sect. 8.2.1. The dataset obtained after applying the jet data quality requirements corresponds to an integrated luminosity of $(15.6 \pm 1.7)\ \mathrm{nb}^{-1}$. For these data, approximately 99% of the calorimeter system was usable for event reconstruction. Details about the jet reconstruction and calibration can be found in Sect. 8.1.

The event selection procedure has been discussed in Sect. 8.2.2. For this study we will use the $L1\_J5$ trigger, which triggers on leading jets with a transverse momentum above 5 GeV at the EM scale. The trigger plateau for this trigger—incorporating the jet energy scale uncertainty—starts at 60 GeV [1], and the minimum mass that can be reconstructed without causing a trigger bias is 340 GeV (see Eq. 10.7).

N. Boelaert, *Dijet Angular Distributions in Proton–Proton Collisions*, Springer Theses,
DOI: 10.1007/978-3-642-24597-8_11, © Springer-Verlag Berlin Heidelberg 2012

Due to the fact that the jet energy scale has large uncertainties for $|\eta| > 2.8$, all jets with $|\eta| > 2.8$ are disregarded. The leading and subleading jets are selected from the remaining jets, and trigger effects are avoided by requiring the transverse momentum of the leading jet to be above 60 GeV. Furthermore the transverse momentum of the subleading jet is required to be above 30 GeV.

Jet selection criteria have been discussed in Sect. 8.2.3. For this analysis we disregard the event if the leading or subleading jet is labeled as *bad*, or if the event contains an *ugly* jet with a transverse momentum above 15 GeV, since these jets can influence the subleading jet.

11.3 Monte Carlo Samples

Monte Carlo samples have been produced with PYTHIA 6.421 using the ATLAS MC09 parameter tune [2]. The samples are QCD dijet events, generated in bins of the transverse momentum of the partons produced by the $2 \rightarrow 2$ matrix elements. This is necessary in order to obtain a reasonable number of events at high p_T.The mass bins in partonic p_T are: [8, 17] GeV, [17, 35] GeV, [35, 70] GeV, [70, 140] GeV, [140, 280] GeV, [280, 560] GeV and [560, 1120] GeV.

The samples have been combined by weighting the events in each sample according to the cross section divided by the total number of events in the sample. The generated events have been sent through a Geant4 simulation of the ATLAS detector [3] and have been reconstructed and analyzed using the same procedure as is applied to data (see Sect. 8.1 for details about jet reconstruction and calibration).

11.4 Physics Selection Cuts

The selection cuts aimed at studying the physics are made in true rapidity and dijet invariant mass. The mass bins were selected to be $340 < M_{jj} < 520$ GeV and $520 < M_{jj} < 680$ GeV.

The rapidity cuts are the following:

$$\begin{aligned} |y_1 + y_2| &< 1.5, \\ |y_1 - y_2| &< \log(30) \approx 3.40 \end{aligned} \qquad (11.2)$$

These cuts differ from the ones specified in Eq. 4.22 due to the use of rapidity instead of pseudorapidity. From a physics point of view, rapidity-difference is preferred to pseudorapidity-difference, since it is invariant under boosts along the longitudinal axis, meaning that we can safely perform the measurement in the laboratory frame. But rapidity is defined using the jet mass, which is a quantity—historically—not well measured (see Sect. 4.22 for the definition of rapidity). Progress in the

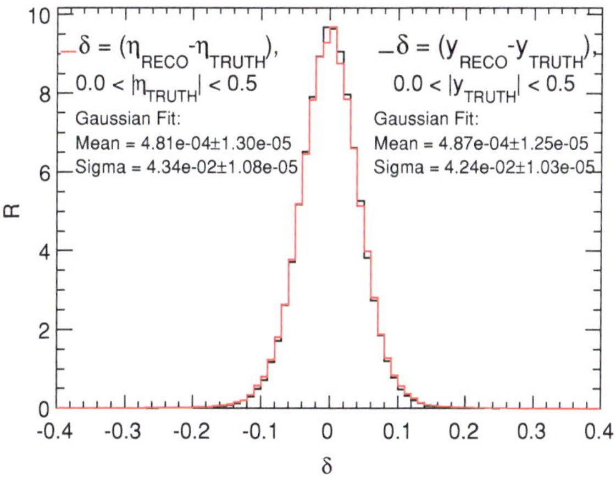

Fig. 11.1 Angular resolutions in rapidity (*black* line) and pseudorapidity (*red* line) for jets for which the true value of $|y|$ (for the rapidity resolution) or $|\eta|$ (for the pseudorapidity resolution) lies between 0 and 0.5. The curves are normalized to unit area

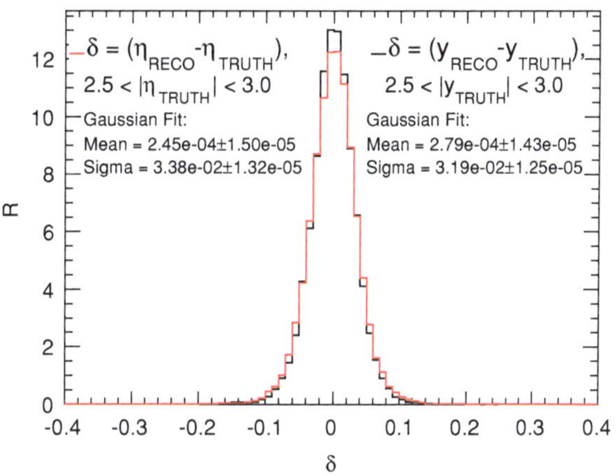

Fig. 11.2 Angular resolutions in rapidity (*black* line) and pseudorapidity (*red* line) for jets for which the true value of $|y|$ (for the rapidity resolution) or $|\eta|$ (for the pseudorapidity resolution) lies between 2.5 and 3. The curves are normalized to unit area

ATLAS jet reconstruction and calibration has improved the reconstruction of rapidity, which is no longer inferior to the reconstruction of pseudorapidity.

This can be understood from comparing the resolution in rapidity with the resolution in pseudorapidity. For this we use the Monte Carlo samples that have been discussed in 11.3 and compare the reconstructed (pseudo)-rapidity value with its underlying truth value. We only consider jets with a transverse momentum above 30 GeV.

Figure 11.1 shows the angular resolution, defined as the difference between the reconstructed and its truth value, in rapidity (black line) and pseudorapidity (red line) for jets for which $|y|$ (for the rapidity resolution) or $|\eta|$ (for the pseudorapidity resolution) lies between 0 and 0.5. The curves are normalized to unit area. The same is shown in Fig. 11.2 for jets with $|y|$ or $|\eta|$ between 2.5 and 3.0. There is no significant difference between the resolution in rapidity and the resolution in pseudorapidity.

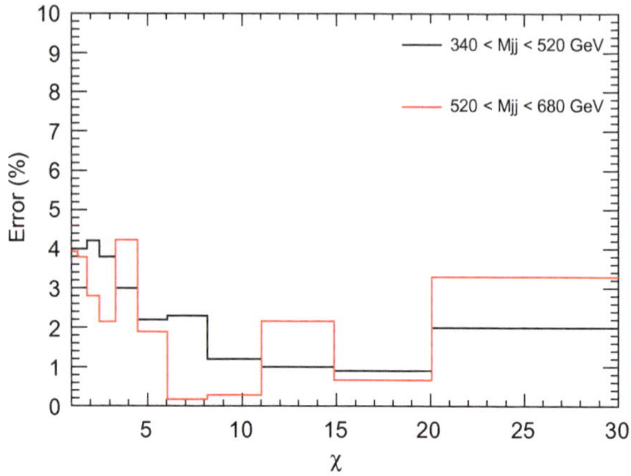

Fig. 11.3 Jet energy scale uncertainties for the mass bins $340 < M_{jj} < 520\,\text{GeV}$ and $520 < M_{jj} < 680\,\text{GeV}$

11.5 Results

11.5.1 Systematic Uncertainties

We have considered the following experimental uncertainties: the uncertainty on the jet energy scale, the jet energy resolution and the angular resolution (see also Sect. 10.6). The major experimental systematic uncertainty on the normalized angular distributions comes from the uncertainty on the jet energy scale (see e.g. Fig. 10.15).

The jet energy scale uncertainty in early data was measured to be $\Delta p_T / p_T < 10\%$ for jets up to $|\eta| < 2.8$ (see Sect. 8.3.1). This uncertainty is assumed to have a Gaussian distribution, with a width $\sigma(p_T, \eta)$ dependent on the transverse momentum and the pseudorapidity of the jet (see Fig. 8.1).

In order to estimate the uncertainty on the dijet angular distributions coming from the uncertainty on the jet scale, we make a large collection of replicated datasets that have been modified to imitate the effect of wrongly calibrated jets; for a certain replicated dataset, the transverse momentum of all jets in all events has been scaled with a factor which is generated from a Gaussian distribution with zero mean and unit standard deviation, multiplied with $\sigma(p_T, \eta)$. We then reconstruct the dijet angular distributions using this modified dataset.

This procedure is repeated many times (of the order of 10^3), and the bin-by-bin comparison between the modified and the nominal—undisturbed—distributions allows for an estimate of the uncertainty by determining the area that contains 68% of the modified distributions, corresponding to a $1-\sigma$ deviation. For the mass bins considered in this chapter, this uncertainty does not exceed 5%, as shown in Fig. 11.3.

The theoretical uncertainties coming from the PDF uncertainties and the scale uncertainties have been determined according to the methods described in Chap. 10,

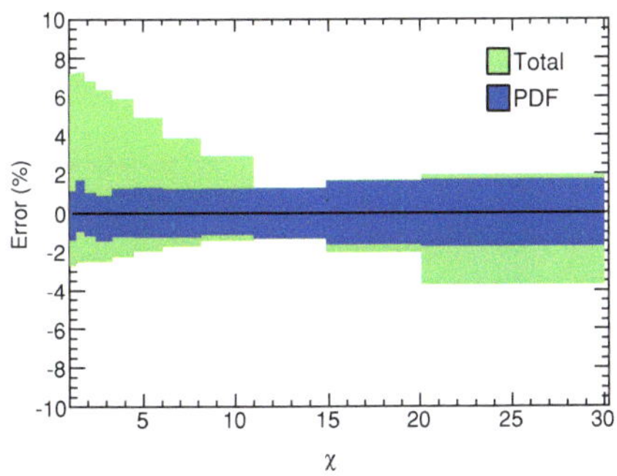

Fig. 11.4 Theoretical uncertainties coming from the PDFs and the scale uncertainties. *Left* mass bin $340 < M_{jj} < 520\,\text{GeV}$, *right* mass bin $520 < M_{jj} < 680\,\text{GeV}$. See the text for a discussion about the uncertainties

and they are summarized in Fig. 11.4 for the mass bins $340 < M_{jj} < 520\,\text{GeV}$ (left figure) and $520 < M_{jj} < 680\,\text{GeV}$ (right figure). The green area shows the total uncertainty calculated as the addition in quadrature of PDF and scale uncertainties, while the blue area shows the PDF uncertainty only. Positive and negative errors have been considered separately; for the ease of representation, they are shown as positive and negative percentages plotted around zero.

11.5.2 Distributions

Figure 11.5 shows the measured normalized dijet angular distributions in the mass bins $340 < M_{jj} < 520\,\text{GeV}$ (top figure) and $520 < M_{jj} < 680\,\text{GeV}$ (bottom figure), together with the Monte Carlo predictions. The lower mass bin contains 1374 events, while the higher mass bin contains 162 events. For these histograms, the data quality cuts discussed in Sect. 11.2 remove $< 2.1\%$ of the events.

The errorbars on the experimental results combine the systematic uncertainty coming from the jet energy scale and the statistical uncertainty in quadrature.

The Monte Carlo prediction has been calculated from the samples discussed in Sect. 11.3. In order to account for NLO effects, the distributions have been multiplied with the k-factors, which have been derived using the method described in Sect. 10.4. The errorbars correspond to the addition in quadrature of PDF and scale uncertainties as well as statistical uncertainties (which are negligible).

Taking into account only statistical uncertainties, we use Eq. A.4. to calculate the ratio of the χ^2 value over the number of degrees of freedom, i.e. the number of histogram bins minus one (since the distributions are normalized), for the lowest mass bin. The result equals 0.60, which implies a good agreement between Monte Carlo prediction and data.

No data unfolding techniques have been applied to correct the measured distributions for detector effects. Even though Monte Carlo studies (see Sect. 10.7) show that

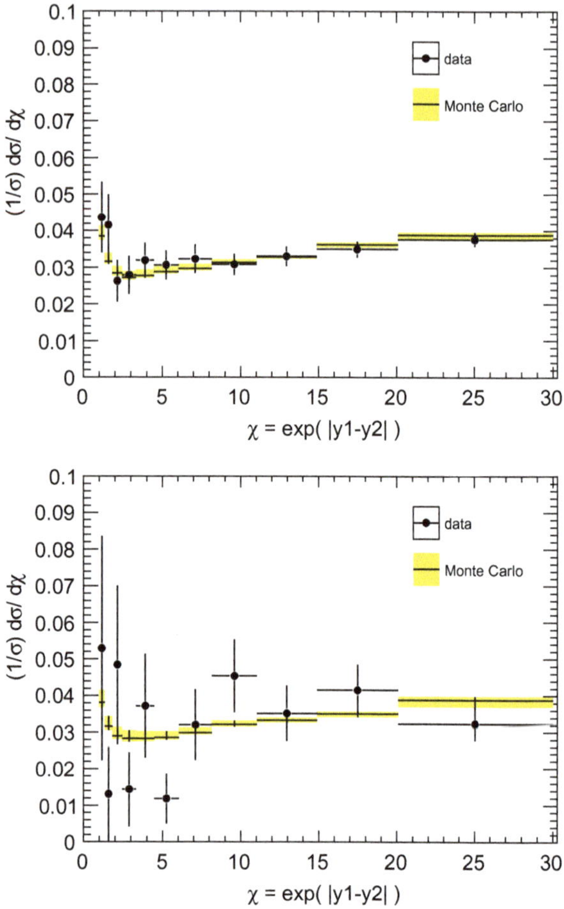

Fig. 11.5 Dijet angular distributions measured by ATLAS, together with the Monte Carlo predictions. Top: mass bin $340 < M_{jj} < 520\,\text{GeV}$, bottom: mass bin $520 < M_{jj} < 680\,\text{GeV}$. See the text for a discussion about the uncertainties

the corrections are at a controllable level, more data and study are needed in order to be able to apply correction factors that do not suffer from large uncertainties.

References

1. ATLAS Collaboration, Measurement of jet production in proton–proton collisions at 7 TeV centre-of-mass energy with the ATLAS Detector. Technical report ATL-COM-PHYS-2010-408, CERN, Geneva (2010)
2. ATLAS Collaboration, ATLAS Monte Carlo tunes for MC09. Technical report ATL-PHYS-PUB-2010-002, CERN, Geneva (2010)
3. Geant4 Collaboration Geant4–a simulation toolkit. Nuclear instruments and methods in physics research section A: accelerators, spectrometers, detectors and associated equipment, **506**(3), 250–303 (2003)

Chapter 12
Conclusions and Outlook

12.1 Conclusions

In the first half of 2010 the LHC had its first proton–proton collisions at $\sqrt{s} = 7$ TeV, which allows the LHC experiments to start investigating a kinematic regime that has never been explored before; the multi TeV range is expected to reveal both Standard Model phenomena and new physics discoveries.

Many new physics models exist, but in this thesis we have chosen to focus on gravity mediated effects coming from large extra dimensions, in addition to the Standard Model. More precisely, we have studied gravitational scattering and black holes in the ADD model where the four-dimensional world membrane has a small but finite width.

Jet production is the most dominant hard process in hadron collision experiments. Jets form the background to many new physics searches, yet they can also be used as a signal, both for probing QCD and for physics beyond the Standard Model. Because of their rich abundance, many jet studies can be performed with little integrated luminosity.

The ATLAS detector is one of the multi-purpose detectors that operate at the LHC. Its hermetic calorimeter system allows measuring jets rather precisely up to high values of pseudorapidity ($\eta_{\rm jet} \sim 4$). First measurements have shown that for early ATLAS data a jet energy scale uncertainty of less than 10 % is obtained for jets up to $|\eta| < 2.8$.

The dijet angular distribution is the differential cross section $d\sigma / d\chi$ versus χ in bins of dijet invariant mass. Dijet angular distributions are an excellent tool to test both QCD and new physics, since they mainly probe the hard matrix element, and do not rely as much as other dijet observables on the jet energy scale.

This thesis has studied dijet angular distributions at $\sqrt{s} = 7$ TeV and $\sqrt{s} = 14$ TeV, the latter one being the LHC design energy which will only be reached after a few years of operating the LHC at $\sqrt{s} = 7$ TeV followed by a long period of shutdown needed to train the magnets for higher energies.

N. Boelaert, *Dijet Angular Distributions in Proton–Proton Collisions*, Springer Theses, DOI: 10.1007/978-3-642-24597-8_12, © Springer-Verlag Berlin Heidelberg 2012

In order to explore the ultimate physics potential of the LHC, a detailed phenomenology study at $\sqrt{s} = 14$ TeV has been carried out to study several aspects of QCD and new physics coming from gravity mediated effects in large extra dimensions.

The conclusion of this study is that for dijet masses in the TeV range and low values of χ, i.e. high values of dijet transverse momentum, a clear discrimination between the rather flat QCD prediction—characterized by mainly gluon t-channel exchanges—and the peaked contribution arising from new physics processes—mainly s-channel exchanges—can be made. At high χ, theoretical uncertainties, such as the uncertainties coming from the choice of the renormalization and factorization scales, start to become important.

The same conclusions can be made for dijet angular distributions at $\sqrt{s} = 7$ TeV. A dedicated Monte Carlo study aimed at preparing ATLAS for the early measurements has been carried out in order to investigate the phenomenology and detector effects.

It has been shown that normalizing the distributions reduces both the theoretical and experimental uncertainty. The sensitivity of ATLAS to gravitational effects coming from large extra dimensions has been investigated, and already with 1 pb^{-1} of data the ATLAS experiment is expected to be sensitive to these effects in the dijet mass bins above 1 TeV. No new physics is expected to show up in the lower dijet mass bins but, given the fact that the systematic uncertainties are well constrained and that by construction the dijet angular distributions mainly probe the hard matrix elements and not so much the convolution of the hard cross section with the PDFs, the measurements in the lower mass bins can be tested against QCD predictions.

We have made the data-QCD comparison using ATLAS data; the normalized dijet angular distributions have been measured by the ATLAS detector in two dijet mass bins, up to $M_{jj} < 680$ GeV, using data that were recorded during April and May of 2010, corresponding to an integrated luminosity of about 15 nb^{-1}. The measurement shows good agreement with QCD predictions.

12.2 Outlook

This thesis is a first look at dijet angular distributions at LHC energies. Only masses below the TeV range that were measured with ATLAS have been analyzed, and no discrepancy with QCD has been found. But with more data being recorded daily, we hope to study higher jet masses very soon. Furthermore, the performance of the ATLAS detector is expected to increase as well, so not only statistical but also experimental uncertainties will decrease rather quickly. This means that at lower dijet masses we will be capable of studying QCD more precisely, while at higher masses we will soon be able to either observe signals that arise from new physics or—if nature turns out to be more complicated—set limits on the new physics models. The LHC era has only just begun.

Appendix A
Statistical Hypothesis Testing Using the Frequentist Method

This appendix describes how to test the hypothesis that the measured dijet angular distribution agrees with a theoretical prediction, which is either the Standard Model or a new physics model. This hypothesis test is done by defining a χ^2 test variable. In the next section we derive this variable in the absence of systematic uncertainties in the data, and we describe how to test the null hypothesis of the Standard Model (no new physics). Since measurements are prone to systematic errors, we generalize in Sect. A.2 the method to account for them. Finally Sect. A.3 describes how to set limits on new physics models. The method described in this appendix is referred to as the *frequentist hypothesis test* and is explained in [1].

A.1 Null Hypothesis Testing Using Method of Maximum Likelihood

We assume that Poisson statistics apply on the dijet distributions. We have k bins labeled by the index i running from 1 to k. We let n_i be the number of events in the ith bin, with $\mathbf{n} = (n_1, n_2, \ldots, n_k)$ and $N = \sum_i n_i$.

Without systematic uncertainties, the mean number of entries is predicted by

$$E[n_i] = \mu_i s_i + b_i, \tag{A.1}$$

with μ_i the strength parameter ($\mu_i = 1$ for the expected signal), b_i the theoretical prediction of the background and s_i the prediction of the signal. Furthermore, we use the notation that $\boldsymbol{\mu} = (\mu_1, \mu_2, \ldots, \mu_k)$. Our task is to test the hypothesis that $\boldsymbol{\mu} = 0$, which can be done using the method of maximum likelihood ratios. For Poisson distributed data, the likelihood L is given by the product of Poisson distributions for each bin:

$$L(\boldsymbol{\mu} : \mathbf{n}) = \prod_i \frac{[(\mu_i s_i + b_i)]^{n_i} \exp[-(\mu_i s_i + b_i)]}{n_i!}, \tag{A.2}$$

DOI: 10.1007/978-3-642-24597-8,

From this we can derive the likelihood ratio $\lambda(\boldsymbol{\mu})$:

$$\lambda(\boldsymbol{\mu}) = \frac{L(\boldsymbol{\mu} : \mathbf{n})}{L(\hat{\boldsymbol{\mu}} : \mathbf{n})}, \tag{A.3}$$

with $\hat{\boldsymbol{\mu}} = (\hat{\mu_1}, \hat{\mu_2}, \ldots, \hat{\mu_k})$ the values of μ_i that maximize the likelihood. Maximizing the likelihood gives the best estimate for the parameters μ_i, and is equivalent to maximizing the likelihood ratio or to minimizing $\chi^2(\boldsymbol{\mu})$ defined by

$$\chi^2(\boldsymbol{\mu}) = -2\ln\lambda = -2\ln L(\boldsymbol{\mu} : \mathbf{n}) + 2\ln L(\hat{\boldsymbol{\mu}} : \mathbf{n}) \tag{A.4}$$

The likelihood ratio test theorem says that χ^2 asymptotically obeys a chi-square distribution with the number of degrees of freedom equal to the independent number of parameters being tested, i.e. $(k-1)$ [1]. Note that the second term of the right hand side of Eq. A.4 is independent of $\boldsymbol{\mu}$ so that the minimization of χ^2 is entirely equivalent to the maximization of the likelihood function L.

For Poisson distributed data, we may replace the unknown values of $\hat{\mu}_i$ by their bin-by-bin model-independent maximum likelihood estimations, which can be easily found by taking the derivatives of $L(\boldsymbol{\mu} : \mathbf{n})$ with respect to μ_i, and solving them for μ_i, giving:

$$n_i s_i [(\mu_i s_i + b_i)]^{n_i - 1} \frac{\exp[-(\mu_i s_i + b_i)]}{n_i!} - s_i [(\mu_i s_i + b_i)]^{n_i} \frac{\exp[-(\mu_i s_i + b_i)]}{n_i!} = 0 \tag{A.5}$$

$$\Longleftrightarrow n_i - (\mu_i s_i + b_i) = 0 \tag{A.6}$$

So that:

$$\hat{\mu}_i = (n_i - b_i)/s_i \tag{A.7}$$

Suppose we want to test the null hypothesis H_0 that $\boldsymbol{\mu} = 0$. We can use $\chi^2(\boldsymbol{\mu} = 0)$ (Eq. A.4) as a test statistic to test whether to accept or reject H_0. Inserting Eq. A.7 into Eq. A.4:

$$\begin{aligned} \chi^2(0) &= -2\sum_i \ln\left(\frac{b_i^{n_i}\exp(-b_i)}{n_i!}\right) + 2\sum_i \ln\left(\frac{n_i^{n_i}\exp(-n_i)}{n_i!}\right) \\ &= 2\sum_i (n_i \ln(n_i/b_i) + (b_i - n_i)) \approx \sum_i \frac{(n_i - b_i)^2}{n_i} \end{aligned} \tag{A.8}$$

In the above expression, the following approximation was used:

$$n_i \ln\left(\frac{n_i}{b_i}\right) = -n_i \ln\left(1 + \frac{b_i - n_i}{n_i}\right) \approx n_i \left[\frac{n_i - b_i}{n_i} + \frac{1}{2}\left(\frac{n_i - b_i}{n_i}\right)^2\right] \tag{A.9}$$

We now use $\chi^2(\boldsymbol{\mu} = 0)$ (Eq. A.4) to derive the *p-value*:

$$p = \int_{\chi^2(\mu=0)}^{\infty} f(z; n_d) \mathrm{d}z, \tag{A.10}$$

with $f(z; n_d)$ the probability density function of the chi-square statistic with n_d degrees of freedom [here $n_d = (k-1)$]. The p-value can be understood as the probability, under the assumption of a hypothesis H_0, of obtaining data at least as incompatible with H_0 as is actually observed. Values for p can be found in standard math references.

The p-value can be related to the *significance*, which is defined as the number of standard deviations Z at which a Gaussian random variable of zero mean would give a one-sided tail area equal to the p-value:

$$p = \int_{Z}^{\infty} \frac{1}{\sqrt{2\pi}} e^{-x^2(\mu=0)/2} = 1 - \Phi(Z), \tag{A.11}$$

where Φ is the cumulative distribution for the standard (zero mean, unit variance) Gaussian. Often in HEP, a significance of $Z = 3$ is regarded as evidence, and $Z = 5$ is taken as discovery. These values correspond to p-values of 1.35×10^{-3} and 2.87×10^{-7} respectively.

A.2 Systematic Uncertainties in the Data

Assume now that we have a systematic uncertainty α_i in each bin, which modifies the expectation of the mean in each bin (Eq. A.1):

$$E[n_i] = (\mu_i s_i + b_i)(1 + \alpha_i), \tag{A.12}$$

with α_i typically called the *nuisance parameter*. Assume that α_i is modeled by a Gaussian with mean zero and σ_{α_i}. For example, if the jet energy scale is established with a 5% uncertainty, then $\sigma = 0.05$.

The likelihood is given by:

$$L(\boldsymbol{\mu} : \boldsymbol{\alpha}) = \prod_i \frac{[(\mu_i s_i + b_i)(1+\alpha_i)]^{n_i} \exp[-(\mu_i s_i + b_i)(1+\alpha_i)]}{n_i!} \frac{1}{\sqrt{2\pi}\sigma_{\alpha_i}} \exp\left(-\frac{\alpha_i^2}{2\sigma_{\alpha_i}^2}\right), \tag{A.13}$$

with $\boldsymbol{\alpha} = (\alpha_1, \ldots, \alpha_k)$. This changes the expression for $\chi^2(\boldsymbol{\mu} = 0)$:

$$\chi^2(\boldsymbol{\mu} = 0) = -2 \ln \frac{L(\boldsymbol{\mu} = 0 : \hat{\hat{\boldsymbol{\alpha}}}(\boldsymbol{\mu} = 0))}{L(\hat{\boldsymbol{\mu}} : \hat{\boldsymbol{\alpha}})}, \tag{A.14}$$

with $\hat{\hat{\alpha}}_i(\mu_i)$ the value of α_i that maximizes the likelihood for a given μ_i. The value of $\hat{\mu}_i$ can be found by maximizing the likelihood $L(\boldsymbol{\mu} : \hat{\hat{\boldsymbol{\alpha}}}(\boldsymbol{\mu}))$ and furthermore, $\hat{\alpha}_i = \hat{\hat{\alpha}}_i(\hat{\mu}_i)$. The calculation of $\hat{\hat{\alpha}}_i(\mu_i)$ is done by taking the derivative in terms of α_i:

$$\frac{\partial L(\boldsymbol{\mu} : \boldsymbol{\alpha})}{\partial \alpha_i} = 0 \Leftrightarrow \frac{\partial}{\partial \alpha_i}(1+\alpha_i)^{n_i} \exp\left[-(\mu_i s_i + b_i)(1+\alpha_i) - \frac{\alpha_i^2}{2\sigma_{\alpha_i}^2}\right] = 0, \tag{A.15}$$

giving:

$$\Leftrightarrow n_i - \left(\mu_i s_i + b_i - \frac{\alpha_i}{\sigma_{\alpha_i}^2}\right)(1+\alpha_i) = 0 \tag{A.16}$$

To solve this, we make the approximation that σ_{α_i} and therefore most likely α_i are small, meaning that $1+\alpha_i \approx 1$. This gives:

$$\hat{\hat{\alpha}}_i(\mu_i) = (n_i - (\mu_i s_i + b_i))\sigma_{\alpha_i}^2 \tag{A.17}$$

Using again $1+\alpha_i \approx 1$:

$$L(\boldsymbol{\mu} : \hat{\hat{\boldsymbol{\alpha}}}(\boldsymbol{\mu})) \approx \prod_i \frac{(\mu_i s_i + b_i)^{n_i} \exp\left[-(\mu_i s_i + b_i) - \frac{1}{2}(n_i - (\mu_i s_i + b_i))^2 \sigma_{\alpha_i}^2\right]}{n_i! \sqrt{2\pi}\sigma_{\alpha_i}} \tag{A.18}$$

We now set the derivative of Eq. A.18 with respect to μ_i equal to zero and solve the result for μ_i:

$$\begin{aligned} 0 &= n_i s_i \left[(\mu_i s_i + b_i)\right]^{n_i - 1} \frac{\exp\left[-(\mu_i s_i + b_i) - \frac{1}{2}(n_i - (\mu_i s_i + b_i))^2 \sigma_{\alpha_i}^2\right]}{n_i!} \\ &\quad - \left(s_i - s_i(n_i - (\mu_i s_i + b_i))\sigma_{\alpha_i}^2\right) \\ &\quad \times \left[(\mu_i s_i + b_i)\right]^{n_i} \frac{\exp\left[-(\mu_i s_i + b_i) - \frac{1}{2}(n_i - (\mu_i s_i + b_i))^2 \sigma_{\alpha_i}^2\right]}{n_i!} \\ &\Longleftrightarrow 0 = n_i - (\mu_i s_i + b_i)\left[1 - (n_i - (b_i + \mu_i s_i))\sigma_{\alpha_i}^2\right], \end{aligned} \tag{A.19}$$

giving:

$$\hat{\mu}_i \approx \frac{n_i - b_i}{s_i} \tag{A.20}$$

We can now calculate χ^2 (Eq. A.14):

$$
\begin{aligned}
\chi^2(0) &= -2\sum_i \ln\left(\frac{b_i^{n_i}\exp\left[-b_i - \frac{1}{2}(b_i - n_i)^2\sigma_{\alpha_i}^2\right]}{n_i!}\right) + 2\sum_i \ln\left(\frac{n_i^{n_i}\exp(-n_i)}{n_i!}\right) \\
&= 2\sum_i \left(n_i \ln(n_i/b_i) + (b_i - n_i) + \frac{1}{2}(b_i - n_i)^2\sigma_{\alpha_i}^2\right) \\
&\approx \sum_i \left[\frac{(n_i - b_i)^2}{n_i} + \frac{(b_i - n_i)^2 n_i \sigma_{\alpha_i}^2}{n_i}\right] \\
&\approx \sum_i \frac{(n_i - b_i)^2}{n_i(1 + n_i\sigma_{\alpha_i}^2)} = \sum_i \frac{(n_i - b_i)^2}{(n_i + n_i^2\sigma_{\alpha_i}^2)}
\end{aligned} \tag{A.21}
$$

The interpretation of the above formula is rather straightforward; the presence of a systematic uncertainty, characterized by $\sigma_{\alpha_i}^2$, reduces the χ^2-value so that it becomes more difficult to reject the null hypothesis. The denominator can be interpreted as the addition of the statistical and systematic uncertainties in quadrature.

A.3 Limits

The previous sections describe how to use data to exclude the null hypothesis of the Standard Model. But data can also be used in a different way, namely for the exclusion of new physics models, i.e. for setting limits on new physics model parameters. We then need to test the hypothesis that $\boldsymbol{\mu} = \boldsymbol{1}$. We can derive $\chi^2(1)$ in a similar way as was done for $\chi^2(0)$, and the result is similar to Eq. (A.21) but with b_i replaced by $b_i + s_i$. Once we have calculated the χ^2 value, we can obtain the p-value. For those model parameters for which the p-value is sufficiently small, we can exclude them, i.e. set limits on the theory. A common choice is to let the critical p-value correspond to a significance of $Z = 2$. In that case $p = 0.05$ and we obtain so called 95 % Confidence Level (C. L.) limits.

Reference

C. Amsler et al., Review of particle physics. Phys. Lett. **B667**, 1 (2008)

Curriculum Vitae

Personal Information

Work Address	Division of Experimental High Energy Physics Niels Bohr Institute University of Copenhagen Blegdamsvej 17 2100 Copenhagen, Denmark
Date of Birth	20/09/1983
Place of Birth	Ghent, Belgium
Nationality	Belgian

Education

10/2006–9/2010	**PhD in physics** **Division of Experimental High-Energy Physics** **Lund University, Sweden** Thesis: "Dijet angular distributions in proton proton collisions at $\sqrt{s} = 7\,\text{TeV}$ and $\sqrt{s} = 14\,\text{TeV}$" Supervisor: Prof. Dr. Torsten Åkesson Marie Curie Fellowship for Early Stage Research Training (EFP6) Member of the ATLAS collaboration Additional training, including "Introduction to Teaching and Learning", "Intellectual Property and the Innovation System", "Leadership and Research Management" and "Project Management"

N. Boelaert, *Dijet Angular Distributions in Proton–Proton Collisions,* Springer Theses,
DOI: 10.1007/978-3-642-24597-8,

09/2001–09/2006 **Master of Engineering Physics: option Applied Physics**
Ghent University, Belgium
Thesis: "Experimental and theoretical study of the neutron deficient Cd isotopes"
Supervisor: Prof. Dr. Kris Heyde
Summa cum laude

Prizes

3/2011 **Oscar II:s Stipendium**
PhD thesis award at Lund University

3/2006 **Miniforschungspreis 2006 der Universität zu Köln**
Award for master's project

Employment

10/2010–present **Postdoc at Niels Bohr Institute**
University of Copenhagen, Denmark
Member of the ATLAS collaboration

10/2006–9/2010 **PhD student at the Division of Experimental High-Energy Physics**
Lund University, Sweden
Member of the ATLAS collaboration

07/2005–09/2005 **Traineeship Institut für Kernphysik**
Universität zu Köln, Germany
Lifetime Determination of Excited States in neutron deficient Cd isotopes

07/2004–09/2004 **Traineeship Montena emc sa**
Fribourg, Switzerland
Engineering work in the area of electromagnetic compatibility

Selected Peer-Reviewed Publications

- N. Boelaert, M. Smirnova, K. Heyde and J. Jolie, *Shell model description of the low-lying states of the neutron deficient Cd isotopes*, Phys. Rev. **C75** (2007) 014316
- N. Boelaert, M. Dewald, C. Fransen, J. Jolie, A. Linneman and B. Melon, *Low-spin electromagnetic transition probabilities in* $\{102, 104\}$*Cd*, Phys. Rev. **C75** (2007) 054311
- N. Boelaert and T. Åkesson, *Dijet angular distributions at* $\sqrt{s} = 14\,\text{TeV}$, arXiv:0905.3961 [hep-ph], EPJ **C33**, (2010) 343

- The ATLAS collaboration, *Search for New Particles in Two-Jet Final States in 7 TeV Proton–Proton Collisions with the ATLAS Detector at the LHC*, arXiv:1008.2461 [hep-ex], Phys. Rev. Lett. 105 (2010) 161801
- The ATLAS collaboration, *Search for quark contact interactions in dijet angular distributions in pp collisions at* $\sqrt{s} = 7\,\text{TeV}$ *measured with the ATLAS detector*, arXiv:1009.5069 [hep-ex], Phys. Lett. **B 694** (2011) 327–345
- The ATLAS collaboration, *Measurement of inclusive jet and dijet cross sections in proton–proton collisions at* 7 TeV *centre-of-mass energy with the ATLAS detector*, arXiv:1009.5908 [hep-ex], Eur. Phys. J. **C 71** (2011) 1512
- The ATLAS collaboration, *Search for New Physics in Dijet Mass and Angular Distributions in Collisions at* $\sqrt{s} = 7\,\text{TeV}$ *Measured with the ATLAS Detector*, arXiv:1103.3864 [hep-ex], New J. Phys. **13** (2011) 053044

Other Publications

- N. Boelaert, *Dijet angular distributions at* $\sqrt{s} = 14\,\text{TeV}$, PoS (EPS-HEP 2009) 298
- N. Boelaert et al., *Software design for prompt assessment of time-varying data quality*, ATLAS internal communication, ATL-COM-GEN-2010-002-Geneva: CERN, 2010
- N. Boelaert, *Implementation of the GravADD generator in Athena*, ATLAS internal note, ATL-PHYS-INT-2010-012-Geneva: CERN, 2010
- N. Boelaert et al., *ATLAS sensitivity to contact interactions and large extra dimensions using dijet events at* $\sqrt{s} = 7\,\text{TeV}$, ATLAS internal communication, ATL-COM-PHYS-2010-136 - Geneva: CERN, 2010
- N. Boelaert et al., *High-pT dijet angular distributions in pp interactions at* $\sqrt{s} = 7\,\text{TeV}$ *measured with the ATLAS detector at the LHC*, ATLAS conference note, ATLAS-CONF-2010-074 - Geneva: CERN, 2010
- N. Boelaert et al., *Backup document for ATLAS publication: Search for quark contact interactions in dijet angular distributions in pp collisions at* $\sqrt{s} = 7\,\text{TeV}$ *measured with the ATLAS detector*, ATLAS internal communication, ATL-COM-PHYS-2010-729 - Geneva: CERN, 2010
- H. Bawa, N. Boelaert, R. Buckingham, G. Choudalakis, P. O. Deviveiros, A. Gibson, C. Issever, F. Ruehr, P. Savard and M. Shupe, *Searches for Quark Contact Interactions and Other New Physics in Dijet Angular Distributions in pp Collisions at* $sqrt(s) = 7\,\text{TeV}$ *Measured with the ATLAS Detector at the LHC*, ATLAS internal communication, ATL-COM-PHYS-2011-032 - Geneva: CERN, 2011

MIX
Papier aus verantwortungsvollen Quellen
Paper from responsible sources
FSC® C105338

If you have any concerns about our products,
you can contact us on
ProductSafety@springernature.com

In case Publisher is established outside the EU,
the EU authorized representative is:
Springer Nature Customer Service Center GmbH
Europaplatz 3, 69115 Heidelberg, Germany

Printed by Libri Plureos GmbH
in Hamburg, Germany